Springer Series in Operations Research and Financial Engineering

Series Editors:
Thomas V. Mikosch
Sidney I. Resnick
Stephen M. Robinson

T0135146

For further volumes:
http://www.springer.com/series/3182

Stefan Schäffler

Global Optimization

A Stochastic Approach

 Springer

Stefan Schäffler
Universität der Bundeswehr München
Fakultät für Elektro- und
    Informationstechnik, EIT1
Werner-Heisenberg-Weg 39
Neubiberg, Germany

ISSN 1431-8598
ISBN 978-1-4899-9280-2   ISBN 978-1-4614-3927-1 (eBook)
DOI 10.1007/978-1-4614-3927-1
Springer New York Heidelberg Dordrecht London

Mathematics Subject Classification (2010): 90C26, 90C15, 90C90, 65C30, 65D30, 60H10, 60H35, 60H40

Printed on acid-free paper

Springer is part of Springer Science+Business Media (www.springer.com)

Preface

Global optimization plays an outstanding role in applied mathematics, because a huge number of problems arising in natural sciences, engineering, and economics can be formulated as global optimization problems. It is impossible to overlook the literature dealing with the subject. This is due to the fact that-unlike local optimization-only small and very special classes of global optimization problems have been investigated and solved using a variety of mathematical tools and numerical approximations. In summary, global optimization seems to be a very inhomogeneous discipline of applied mathematics (comparable to the theory of partial differential equations). Furthermore, the more comprehensive the considered class of global optimization problems, the smaller the tractable scale of problems.

In this book, we try to overcome these drawbacks by the development of a homogeneous class of numerical methods for a very large class of global optimization problems. The main idea goes back to 1953, when Metropolis et al. proposed their algorithm for the efficient simulation of the evolution of a solid to thermal equilibrium (see [Met.etal53]).

In [Pin70], the analogy between statistical mechanics and optimization is already noticed for the first time. Since 1985, one tries to use this analogy for solving unconstrained global optimization problems with twice continuously differentiable objective functions (see [Al-Pe.etal85], [GemHwa86], and [Chi.etal87]). These new algorithms are known as *simulated annealing*. Unfortunately, simulated annealing algorithms use so-called cooling strategies inspired by statistical mechanics in order to solve global optimization problems but neglect the existence of efficient local optimization procedures. Hence, these cooling strategies lead to unsatisfactory practical results in general.

The analogy between statistical mechanics and global optimization with constant temperature is analyzed for the first time in [Schä93] using Brownian Motion and using the stability of random dynamical systems. This analysis forms the basis of all methods developed in this book. As a result, the application of the equilibrium theory of statistical mechanics with fixed temperature in combination with the stability theory of random dynamical systems leads to the algorithmic generation of pseudorandom vectors, which are located in the region of attraction of a global optimum point of a given objective function.

Here is an outline of the book. In Chap. 1, stochastic methods in global optimization are summarized. Surveys of deterministic approaches can be found in [Flo00], [HorTui96], and [StrSer00] for instance. In Chap. 2, we develop unconstrained local minimization problems and their numerical analysis using a special type of dynamical systems given by the curve of steepest descent. This approach allows the interpretation of several numerical methods like the Newton method or the trust region method from a unified point of view.

The treatment of global optimization begins with Chap. 3, in which we consider unconstrained global minimization problems. A suitable randomization of the curve of steepest descent by a Brownian Motion yields a class of new non-deterministic algorithms for unconstrained global minimization problems. These algorithms are applicable to a large class of objective functions, and their efficiency does not substantially depend on the dimension of the given optimization problem, which is confirmed by numerical examples. In Chap. 4, we propose a very important application of the results of Chap. 3, namely, the optimal decoding of high-dimensional block codes in digital communications. Chapter 5 is concerned with constrained global minimization problems. Beginning with equality constraints, the projected curve of steepest descent and its randomized counterpart are introduced for local and global optimization, respectively. Furthermore, the penalty approach is analyzed in this context. Besides the application of slack variables, an active set strategy for inequality constraints is developed. The main ideas for global minimization of real-valued objective functions can be generalized to vector-valued objective functions. This is done in the final chapter by the introduction of randomized curves of dominated points.

Appendix A offers a short course in probability theory from a measure-theoretic point of view and Appendix B deals with the algorithmical generation of pseudo-random numbers, which represents the fundament of all numerical investigations in this book. Since we have chosen a stochastic approach for the analysis and numerical solution of global optimization problems, we evidently have to ask whether this approach is adequate when dealing with stochastic global optimization problems. This question is answered in Appendix C by the investigation of gradient information additively disturbed by a white noise process.

The reader of this book should be familiar with

- Initial value problems
- Theory and practice of local optimization
- Topics in probability theory summarized in Appendix A

I gratefully acknowledge the critical comments and suggestions of Prof. K. Pilzweger, Prof. M. Richter, Prof. S. M. Robinson, and Dr. R. von Chossy. Obviously, all errors are my sole responsibility.

I am deeply indebted to the publisher, in particular, to the publisher's representative, Ms. Vaishali Damle, for her patience and understanding.

This book is dedicated to my teachers and promoters Prof. Klaus Ritter and Prof. Stephen M. Robinson.

Notation and Symbols

AWGN	Additive White Gaussian Noise
$\mathcal{B}(\Omega)$	Borel σ-field on Ω
$C^l(M_1, M_2)$	Set of l times continuously differentiable functions $f : M_1 \rightarrow M_2$
$C_{\text{con}}(\bullet)$	Convex hull
$\text{Cov}(\bullet)$	Covariance matrix
δ_{t_0}	Dirac delta function
$\mathbb{E}(\bullet)$	Expectation
f^+	Positive part of f
f^-	Negative part of f
$g.c.d.(\bullet, \bullet)$	Greatest common divisor
globmin	Global minimization
$\mathbf{I}_n$	n-dimensional identity matrix
$I_\bullet$	Indicator function
$L^r\text{-}\lim\limits_{i\to\infty} X_i = X$	$\lim\limits_{i\to\infty} \int |X - X_i|^r \, d\mathbb{P} = 0$
locmin	Local minimization
∇f	Gradient of f
$\nabla^2 f$	Hessian of f
$\mathcal{N}(\mathbf{e}, \boldsymbol{\Sigma})$	n-dimensional Gaussian distribution with expectation $\mathbf{e}$ and covariance matrix $\boldsymbol{\Sigma}$
$\nu_{\mathbf{e},\boldsymbol{\Sigma}}$	Lebesgue density of a $\mathcal{N}(\mathbf{e}, \boldsymbol{\Sigma})$ Gaussian distributed random variable
$\mathcal{P}(\bullet)$	Power set
$\mathbb{P}(\bullet)$	Probability
$\mathbb{P}(\bullet|B)$	Conditional probability
$\mathbb{P}_X$	Image measure of X

q.e.d.	End of proof
SNR	**S**ignal to **N**oise **R**atio
$\sigma(X)$	σ-field generated by X
$\mathbb{V}(\bullet)$	Variance
vlocmin	Local vector minimization
vglobmin	Global vector minimization
$\| \bullet \|_2$	Euclidean norm
$\int_0^T \mathbf{Y}_t \circ d\mathbf{B}_t$	Fisk–Stratonovich integral
$\mathbf{u} \leq_P \mathbf{v}$	$u_i \leq v_i, \quad i = 1, \dots, n, \quad \mathbf{u}, \mathbf{v} \in \mathbb{R}^n$
$\oplus$	Binary addition
$\boxplus$	Addition modulo
$\odot$	Binary multiplication
$\boxdot$	Multiplication modulo
$(\Omega, \mathcal{S})$	Measurable space
$(\Omega, \mathcal{S}, \mu)$	Measure space
$(\Omega, \mathcal{S}, \mathbb{P})$	Probability space
$a \equiv_m b$	a is congruent to b modulo m
$\mathbb{Z}/m\mathbb{Z}$	$\{[0], \dots, [m-1]\}$

Contents

List of Figures

Chapter 1
Stochastic Approach to Global Optimization at a Glance

1.1 Random Search

Let

$$f : R(\subseteq \mathbb{R}^n) \to \mathbb{R}$$

be a given objective function of a global minimization problem and let P be a probability distribution defined on the feasible region R, then a random search algorithm is defined as follows:

1. Compute m pseudorandom vectors $\mathbf{x}_1, \ldots, \mathbf{x}_m$ as realizations of m stochastically independent identically P-distributed random variables

$$\mathbf{X}_1, \ldots, \mathbf{X}_m.$$

2. Choose $j \in \{1, \ldots, m\}$ such that

$$f(\mathbf{x}_j) \le f(\mathbf{x}_i) \quad \text{for all} \quad i \in \{1, \ldots, m\}.$$

3. Use $\mathbf{x}_j$ as global minimum point or apply a local minimization procedure with starting point $\mathbf{x}_j$ to minimize f.

The utility of this algorithm depends mainly on the choice of the probability distribution P as the following example shows.

Example 1.1. Choose $n \in \mathbb{N}$ and consider the function

$$f_n : [-1, 1]^n \to \mathbb{R}, \quad \mathbf{x} \mapsto \sum_{i=1}^{n} (4x_i^2 - \cos(8x_i) + 1).$$

Each function f_n has 3^n isolated minimum points with a unique global minimum point at $\mathbf{x} = \mathbf{0}$ (Figs. 1.1 and 1.2).

S. Schäffler, *Global Optimization: A Stochastic Approach*, Springer Series in Operations Research and Financial Engineering, DOI 10.1007/978-1-4614-3927-1_1, © Springer Science+Business Media New York 2012

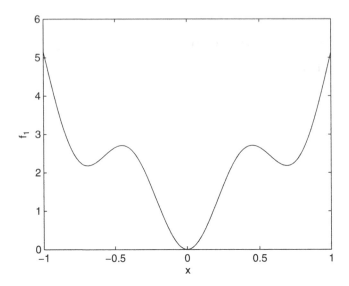

Fig. 1.1 Function f_1

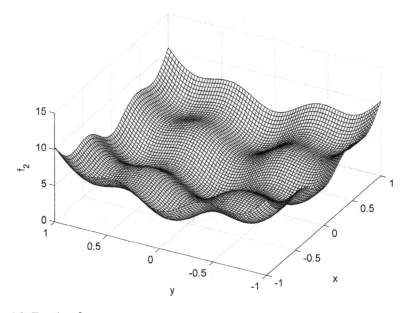

Fig. 1.2 Function f_2

The region of attraction to the unique global minimum point is approximately given by $[-0.4, 0.4]^n$.

Using the uniform probability distribution $P_{u,n}$ on $[-1, 1]^n$ for $\mathbf{X}_1, \ldots, \mathbf{X}_m$, then

$$p_{u,n} := P_{u,n}(\mathbf{X}_i \in [-0.4, 0.4]^n) = 0.4^n$$

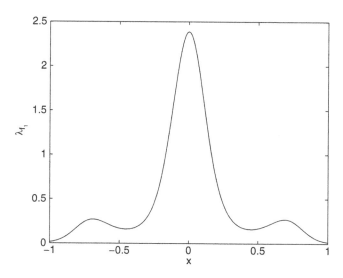

Fig. 1.3 Lebesgue density function λ_{f_1}

denotes the probability that a realization of $\mathbf{X}_i$ lies in the region of attraction of the global minimum point.

Using the probability distribution P_{f_n} given by the Lebesgue density function

$$\lambda_{f_n} : [-1, 1]^n \to \mathbb{R}, \quad \mathbf{x} \mapsto \frac{\exp\left(-f_n(\mathbf{x})\right)}{\int\limits_{[-1,1]^n} \exp\left(-f_n(\mathbf{x})\right) d\mathbf{x}}$$

for $\mathbf{X}_1, \ldots, \mathbf{X}_m$, then

$$p_{f_n} := \frac{\int\limits_{[-0.4,0.4]^n} \exp\left(-f_n(\mathbf{x})\right) d\mathbf{x}}{\int\limits_{[-1,1]^n} \exp\left(-f_n(\mathbf{x})\right) d\mathbf{x}} \approx 0.8^n$$

again denotes the probability that a realization of $\mathbf{X}_i$ lies in the region of attraction of the global minimum point (Figs. 1.3 and 1.4).

We obtain

$$p_{f_n} \approx 2^n p_{u,n}.$$

Clearly, the efficiency of pure random search depends mainly on the choice of P.

In order to make use of the advantage of P_{f_n} over $P_{u,n}$, one has to find a suitable Lebesgue density function λ_f for a given objective function f and one has to develop an algorithm for the generation of realizations according to λ_f. Both will be done in the next chapters. Random search algorithms are investigated in [HenTót10] and [ZhiŽil08] and analyzed in [Kar63] and [Zhi91] using extreme-order statistics.

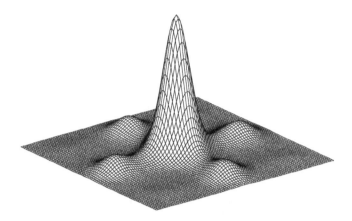

Fig. 1.4 Lebesgue density function λ_{f_2}

1.2 Adaptive Search

Adaptive search algorithms for the global minimization of

$$f : R(\subseteq \mathbb{R}^n) \to \mathbb{R}$$

do the following:

1. Compute a pseudorandom vector $\mathbf{x}_1$ as a realization of a P-distributed random variable, where P denotes a probability measure on R.
2. Based on k computed vectors $\mathbf{x}_1, \ldots, \mathbf{x}_k$, compute a pseudorandom vector $\mathbf{x}_{k+1}$ as a realization of a P_k-distributed random variable, where P_k denotes a probability measure on the set

$$\{\mathbf{x} \in R; \ f(\mathbf{x}) < f(\mathbf{x}_i), \ i = 1, \ldots, k\}.$$

3. Check some termination condition.

Obviously, adaptive search algorithms generate a strictly monotonic decreasing sequence $\{f(\mathbf{x}_i)\}$ of function values at random. In the last 25 years, there has been a great amount of activity related to adaptive search algorithms (see, e.g., [Bul.etal03], [Pat.etal89], [WoodZab02], and [Zab03]). Nevertheless, it has turned out that the computation of a pseudorandom vector in

$$\{\mathbf{x} \in R; \ f(\mathbf{x}) < f(\mathbf{x}_i), \ i = 1, \ldots, k\}$$

by an appropriate probability measure P_k is too difficult in general.

1.3 Markovian Algorithms

A general Markovian Algorithm for the global minimization of

$$f : R(\subseteq \mathbb{R}^n) \to \mathbb{R}$$

can be formulated as follows (cf. [ZhiŽil08]):

(i) By sampling from a given probability distribution P_1 on R, obtain a pseudo-random vector $\mathbf{x}_1$. Evaluate $y_1 = f(\mathbf{x}_1)$ and set iteration number $k = 1$.

(ii) Obtain a pseudorandom vector $\mathbf{z}_k \in R$ by sampling from a probability distribution $Q_k(\mathbf{x}_k)$ on R, which may depend on k and $\mathbf{x}_k$.

(iii) Evaluate $f(\mathbf{z}_k)$ and set

$$\mathbf{x}_{k+1} = \begin{cases} \mathbf{z}_k & \text{with probability} \quad p_k \\ \mathbf{x}_k & \text{with probability} \quad 1 - p_k \end{cases},$$

where p_k is the so-called acceptance probability, which may depend on

$$\mathbf{x}_k, \mathbf{z}_k, f(\mathbf{z}_k), \text{ and } y_k.$$

(iv) Set

$$y_{k+1} = \begin{cases} f(\mathbf{z}_k) & \text{if} \quad \mathbf{x}_{k+1} = \mathbf{z}_k \\ y_k & \text{if} \quad \mathbf{x}_{k+1} = \mathbf{x}_k \end{cases}.$$

(v) Check a stopping criterion. If the algorithm does not stop, substitute $k + 1$ for k and return to Step (ii).

The best-known Markovian algorithm is *simulated annealing*. Let

$$M^* := \{\mathbf{x}^* \in R; \mathbf{x}^* \text{ is a global minimum point of } f\} \neq \emptyset$$

be the set of all global minimum points of f and let P_{M^*} be a probability distribution on R such that

$$P_{M^*}(M^*) = 1,$$

then a simulated annealing algorithm tries to approximate a pseudorandom vector $\mathbf{w}^*$ according to P_{M^*} (in other words $\mathbf{w}^* \in M^*$ almost surely). The acceptance probability is chosen by

$$p_k = \min\{1, \exp(-\beta_k(f(\mathbf{z}_k) - f(\mathbf{x}_k)))\} =$$

$$= \begin{cases} 1 & \text{if} \quad f(\mathbf{z}_k) - f(\mathbf{x}_k) \leq 0 \\ \exp(-\beta_k(f(\mathbf{z}_k) - f(\mathbf{x}_k))) & \text{if} \quad f(\mathbf{z}_k) - f(\mathbf{x}_k) > 0 \end{cases},$$

with $\beta_k \geq 0$ for all $k \in \mathbb{N}$. In order to achieve an approximation of a pseudorandom vector $\mathbf{w}^* \in M^*$, one has to choose the sequence $\{\beta_k\}$ such that

$$\lim_{k \to \infty} \beta_k = \infty \quad \text{(a so-called cooling strategy)}.$$

The efficiency and the theoretical properties of simulated annealing depend mainly on the choice of the probability distributions $Q_k(\mathbf{x}_k)$. Unfortunately, it is not easy to find an optimal class of probability distributions $Q_k(\mathbf{x}_k)$ for a given global optimization problem. Furthermore, there are additional restrictions for the sequence $\{\beta_k\}$ (see [Hajek88] and [Mitra.etal86] for instance) depending on (unknown) properties of the objective function to ensure the approximation of

$$\mathbf{w}^* \in M^* \quad \text{almost surely}.$$

Therefore, many heuristic arguments have been produced (in a highly visible number of publications) to improve simulated annealing (see, e.g., [Sal.etal02] and [LaaAarts87]).

1.4 Population Algorithms

Population algorithms keep a set (the so-called population) of feasible points as a base to generate new points (by random). The set of points evolves by occasionally replacing old by newly generated members according to function values. After the publication of the book by Holland [Holl75] a special type of population algorithms became popular under the name Genetic Algorithms. Similar population algorithms became known under the names of evolutionary programming, genetic programming, and memetic programming. The lack of theoretical foundations and consequently of theoretical analysis of this type of algorithms is usually compensated by an exuberant creativity in finding terms from biology like evolution, genotype, selection, reproduction, recombination, chromosomes, survivor, parents, and descendants. The fact that these algorithms are still very popular is caused by this practice. Population algorithms are characterized by a large number of parameters in general: The MATLAB Genetic Algorithm, for instance, can be adjusted by setting 26 parameters.

In 1995, the biological terminology was enlarged by names like "swarming intelligence" and "cognitive consistency" from behavior research (see [KenEber95]) without improvement of the methods. The new catch phrase is "artificial life." A short and apposite survey of population algorithms is given in [HenTót10].

Chapter 2
Unconstrained Local Optimization

2.1 The Curve of Steepest Descent

In this chapter, we investigate unconstrained local minimization problems of the following type:

$$\mathop{\text{locmin}}_{\mathbf{x}}\{f(\mathbf{x})\}, \quad f : \mathbb{R}^n \to \mathbb{R}, \ n \in \mathbb{N}, \quad f \in C^2(\mathbb{R}^n, \mathbb{R}),$$

where $C^l(\mathbb{R}^n, \mathbb{R})$ denotes the set of l times continuously differentiable functions $g : \mathbb{R}^n \to \mathbb{R}$ ($l = 0$: continuous functions) and where f is called objective function. Hence, we have to compute a point $\mathbf{x}_{\text{loc}} \in \mathbb{R}^n$ such that

$$f(\mathbf{x}) \geq f(\mathbf{x}_{\text{loc}}) \quad \text{for all } \mathbf{x} \in U(\mathbf{x}_{\text{loc}}),$$

where $U(\mathbf{x}_{\text{loc}}) \subseteq \mathbb{R}^n$ is an open neighborhood of $\mathbf{x}_{\text{loc}}$.
The curve of steepest descent

$$\mathbf{x} : \mathbb{D} \subseteq [0, \infty) \to \mathbb{R}^n$$

corresponding to this unconstrained local minimization problem with starting point $\mathbf{x}_0 \in \mathbb{R}^n$ is defined by the solution of the initial value problem

$$\dot{\mathbf{x}}(t) = -\nabla f(\mathbf{x}(t)), \quad \mathbf{x}(0) = \mathbf{x}_0,$$

where $\nabla f : \mathbb{R}^n \to \mathbb{R}^n$ denotes the gradient of the objective function f.
In the following theorem, we summarize some important properties of the above initial value problem.

S. Schäffler, *Global Optimization: A Stochastic Approach*, Springer Series in Operations Research and Financial Engineering, DOI 10.1007/978-1-4614-3927-1_2, © Springer Science+Business Media New York 2012

Theorem 2.1 *Consider*

$$f : \mathbb{R}^n \to \mathbb{R}, \ n \in \mathbb{N}, \quad f \in C^2(\mathbb{R}^n, \mathbb{R}), \quad \mathbf{x}_0 \in \mathbb{R}^n,$$

and let the set

$$L_{f,\mathbf{x}_0} := \{ \mathbf{x} \in \mathbb{R}^n; \ f(\mathbf{x}) \le f(\mathbf{x}_0) \}$$

be bounded, then we have the following:

(i) *The initial value problem*

$$\dot{\mathbf{x}}(t) = -\nabla f(\mathbf{x}(t)), \quad \mathbf{x}(0) = \mathbf{x}_0,$$

   has a unique solution $\mathbf{x} : [0, \infty) \to \mathbb{R}^n$.

(ii) *Either*

$$\mathbf{x} \equiv \mathbf{x}_0 \quad \textit{iff} \quad \nabla f(\mathbf{x}_0) = \mathbf{0}$$

  or

$$f(\mathbf{x}(t + h)) < f(\mathbf{x}(t)) \quad \textit{for all} \quad t, h \in [0, \infty), h > 0.$$

(iii) *There exists a point* $\mathbf{x}_{stat} \in \mathbb{R}^n$ *with*

$$\lim_{t \to \infty} f(\mathbf{x}(t)) = f(\mathbf{x}_{stat}) \quad \textit{and} \quad \nabla f(\mathbf{x}_{stat}) = \mathbf{0}.$$

Proof. Since $L_{f,\mathbf{x}_0} = \{ \mathbf{x} \in \mathbb{R}^n; \ f(\mathbf{x}) \le f(\mathbf{x}_0) \}$ is bounded and $f \in C^2(\mathbb{R}^n, \mathbb{R})$, the set $L_{f,\mathbf{x}_0}$ is compact and there exists a $r > 0$ such that

$$\{ \mathbf{x} \in \mathbb{R}^n; \ f(\mathbf{x}) \le f(\mathbf{x}_0) \} \subseteq \{ \mathbf{x} \in \mathbb{R}^n; \ \|\mathbf{x}\|_2 \le r \}.$$

Setting

$$\mathbf{g} : \mathbb{R}^n \to \mathbb{R}^n, \quad \mathbf{x} \mapsto \begin{cases} \nabla f(\mathbf{x}) & \text{if} \quad \|\mathbf{x}\|_2 \le r \\ \nabla f\left(\frac{r\mathbf{x}}{\|\mathbf{x}\|_2} \right) & \text{if} \quad \|\mathbf{x}\|_2 > r \end{cases},$$

we consider the initial value problem

$$\dot{\mathbf{z}}(t) = -\mathbf{g}(\mathbf{z}(t)), \quad \mathbf{z}(0) = \mathbf{x}_0.$$

Since $\mathbf{g}$ is globally Lipschitz continuous with some Lipschitz constant $L > 0$, we can prove existence and uniqueness of a solution $\mathbf{z} : [0, \infty) \to \mathbb{R}^n$ of this initial value problem using the fixed point theorem of Banach. Therefore, we choose any $T > 0$ and investigate the integral form

$$\mathbf{z}(t) = \mathbf{x}_0 - \int_0^t \mathbf{g}(\mathbf{z}(\tau)) d\tau, \quad t \in [0, T].$$

Let $C^0([0, T], \mathbb{R}^n)$ be the set of all continuous functions $\mathbf{u} : [0, T] \to \mathbb{R}^n$ and

$$K : C^0([0, T], \mathbb{R}^n) \to C^0([0, T], \mathbb{R}^n),$$

$$K(\mathbf{u})(t) = \mathbf{x}_0 - \int_0^t \mathbf{g}(\mathbf{u}(\tau)) d\tau, \quad t \in [0, T].$$

Obviously, each solution of the initial value problem

$$\dot{\mathbf{z}}(t) = -\mathbf{g}(\mathbf{z}(t)), \quad \mathbf{z}(0) = \mathbf{x}_0, \quad t \in [0, T]$$

is a fixed point $\mathbf{z}_T$ of K and vice versa. With

$$d : C^0([0, T], \mathbb{R}^n) \times C^0([0, T], \mathbb{R}^n) \to \mathbb{R}, \quad (\mathbf{u}, \mathbf{v}) \mapsto \max_{t \in [0, T]} \left(\|\mathbf{u}(t) - \mathbf{v}(t)\|_2 e^{-2Lt} \right),$$

it is well known that $(C^0([0, T], \mathbb{R}^n), d)$ is a complete metric space.

The calculation

$$\|K(\mathbf{u})(t) - K(\mathbf{v})(t)\|_2 e^{-2Lt} = \left\| \int_0^t (\mathbf{g}(\mathbf{v}(\tau)) - \mathbf{g}(\mathbf{u}(\tau))) d\tau \right\|_2 e^{-2Lt}$$

$$\leq \int_0^t \|\mathbf{g}(\mathbf{v}(\tau)) - \mathbf{g}(\mathbf{u}(\tau))\|_2 d\tau \cdot e^{-2Lt}$$

$$= \int_0^t \|\mathbf{g}(\mathbf{v}(\tau)) - \mathbf{g}(\mathbf{u}(\tau))\|_2 e^{-2L\tau} e^{2L\tau} d\tau \cdot e^{-2Lt}$$

$$\leq L \int_0^t \|\mathbf{v}(\tau) - \mathbf{u}(\tau)\|_2 e^{-2L\tau} e^{2L\tau} d\tau \cdot e^{-2Lt}$$

$$\leq L \cdot d(\mathbf{u}, \mathbf{v}) \int_0^t e^{2L\tau} d\tau \cdot e^{-2Lt}$$

$$= L \cdot d(\mathbf{u}, \mathbf{v}) \frac{1}{2L} \left(e^{2Lt} - 1 \right) e^{-2Lt}$$

$$\leq \frac{L}{2L} d(\mathbf{u}, \mathbf{v}) = \frac{1}{2} d(\mathbf{u}, \mathbf{v}), \quad t \in [0, T]$$

shows that

$$d(K(\mathbf{u}), K(\mathbf{v})) \leq \frac{1}{2} d(\mathbf{u}, \mathbf{v})$$

so that the fixed point theorem of Banach is applicable. Consequently, we have found a unique solution $\mathbf{z}_T : [0, T] \rightarrow \mathbb{R}^n$ of

$$\dot{\mathbf{z}}(t) = -\mathbf{g}(\mathbf{z}(t)), \quad \mathbf{z}(0) = \mathbf{x}_0$$

for all $T > 0$, and this gives us a unique solution $\mathbf{z} : [0, \infty) \rightarrow \mathbb{R}^n$ of

$$\dot{\mathbf{z}}(t) = -\mathbf{g}(\mathbf{z}(t)), \quad \mathbf{z}(0) = \mathbf{x}_0.$$

Now, we assume

$$\nabla f(\mathbf{x}_0) \neq \mathbf{0}$$

(if $\nabla f(\mathbf{x}_0) = \mathbf{0}$, there is nothing to do) and consider the function

$$\dot{f}(\mathbf{z}(\bullet)) : [0, \infty) \rightarrow \mathbb{R}, \quad t \mapsto \frac{d}{dt} f(\mathbf{z}(t)) \quad (= -\nabla f(\mathbf{z}(t))^\top \mathbf{g}(\mathbf{z}(t)))$$

(one-sided differential quotient for $t = 0$). For $t = 0$, we obtain

$$\dot{f}(\mathbf{z}(0)) = -\nabla f(\mathbf{x}(0))^\top \mathbf{g}(\mathbf{x}(0)) = -\nabla f(\mathbf{x}(0))^\top \nabla f(\mathbf{x}(0)) < 0.$$

Since $\dot{f}(\mathbf{z}(\bullet))$ is continuous, either there exists a smallest $\theta > 0$ with

$$\dot{f}(\mathbf{z}(\theta)) = 0$$

or

$$\dot{f}(\mathbf{z}(t)) < 0 \quad \text{for all } t \in [0, \infty).$$

If such a $\theta > 0$ exists, then $\mathbf{z}(t) \in L_{f,\mathbf{x}_0}$ for all $t \in [0, \theta]$, and the initial value problem

$$\dot{\mathbf{w}}(t) = \mathbf{g}(\mathbf{w}(t)) (= \nabla f(\mathbf{w}(t))), \quad \mathbf{w}(0) = \mathbf{z}(\theta), \quad t \in [0, \theta]$$

has to have two different solutions

$$\mathbf{w}_1 : [0, \theta] \rightarrow \mathbb{R}^n, \quad t \mapsto \mathbf{z}(\theta)$$
$$\mathbf{w}_2 : [0, \theta] \rightarrow \mathbb{R}^n, \quad t \mapsto \mathbf{z}(\theta - t)$$

which is a contradiction to the Lipschitz continuity of $\mathbf{g}$. Hence,

$$\dot{f}(\mathbf{z}(t)) < 0 \quad \text{for all } t \in [0, \infty)$$

and therefore,

$$\mathbf{z}(t) \in \{\mathbf{x} \in \mathbb{R}^n; \ f(\mathbf{x}) \leq f(\mathbf{x}_0)\} \quad \text{for all } t \in [0, \infty).$$

Consequently, the unique solution

$$\mathbf{x} : [0, \infty) \rightarrow \mathbb{R}^n$$

of the initial value problem

$$\dot{\mathbf{x}}(t) = -\nabla f(\mathbf{x}(t)), \quad \mathbf{x}(0) = \mathbf{x}_0$$

is given by

$$\mathbf{x} = \mathbf{z}$$

(part (i) of the theorem). From

$$\dot{f}(\mathbf{x}(t)) < 0 \quad \text{for all } t \in [0, \infty),$$

we obtain

$$f(\mathbf{x}(t + h)) < f(\mathbf{x}(t)) \quad \text{for all} \quad t, h \in [0, \infty), h > 0$$

(part(ii) of the theorem).

Since

$$\mathbf{x}(t) \in \{\mathbf{x} \in \mathbb{R}^n;\ f(\mathbf{x}) \le f(\mathbf{x}_0)\} \quad \text{for all } t \in [0, \infty),$$

we get immediately for all $t \in [0, \infty)$

$$f(\mathbf{x}(0)) \ge f(\mathbf{x}(t)) \ge \min_{\mathbf{y} \in \{\mathbf{x} \in \mathbb{R}^n;\ f(\mathbf{x}) \le f(\mathbf{x}_0)\}} \{f(\mathbf{y})\} > -\infty$$

by the fact that

$$\{\mathbf{x} \in \mathbb{R}^n;\ f(\mathbf{x}) \le f(\mathbf{x}_0)\}$$

is compact and that f is continuous. Since $f(\mathbf{x}(t))$ is monotonically decreasing in t and bounded from below, there exists a $M \in \mathbb{R}$ with

$$\lim_{t \to \infty} f(\mathbf{x}(t)) = M.$$

Therefore,

$$M - f(\mathbf{x}_0) = \int_0^\infty \dot{f}(\mathbf{x}(t)) dt = -\int_0^\infty \|\nabla f(\mathbf{x}(t))\|_2^2 dt.$$

From this equation and the fact that

$$\|\nabla f(\mathbf{x}(t))\|_2^2 > 0 \quad \text{for all } t > 0$$

follows the existence of a sequence $\{\mathbf{x}(t_k)\}_{k \in \mathbb{N}}$ with

$$0 \le t_k < t_{k+1}, \; k \in \mathbb{N}, \quad \lim_{k \to \infty} t_k = \infty, \quad \text{and} \; \lim_{k \to \infty} \nabla f(\mathbf{x}(t_k)) = \mathbf{0}.$$

Since

$$\mathbf{x}(t_k) \in \{\mathbf{x} \in \mathbb{R}^n; \; f(\mathbf{x}) \le f(\mathbf{x}_0)\} \quad \text{for all } k \in \mathbb{N},$$

there exists a convergent subsequence $\{\mathbf{x}(t_{k_j})\}_{j \in \mathbb{N}}$ with

$$1 \le k_j < k_{j+1}, \; j \in \mathbb{N}, \quad \lim_{j \to \infty} k_j = \infty, \quad \text{and} \; \lim_{j \to \infty} \mathbf{x}(t_{k_j}) = \mathbf{x}_{\text{stat}}.$$

Finally, we get

$$\lim_{j \to \infty} f(\mathbf{x}(t_{k_j})) = M = f(\mathbf{x}_{\text{stat}}) \quad \text{and} \quad \nabla f(\mathbf{x}_{\text{stat}}) = \mathbf{0}.$$

q.e.d.

The curve of steepest descent given by

$$\dot{\mathbf{x}}(t) = -\nabla f(\mathbf{x}(t)), \quad \mathbf{x}(0) = \mathbf{x}_0$$

is regular because $\|\dot{\mathbf{x}}(t)\|_2 > 0$ for all $t \in [0, \infty)$. Therefore, it is common to consider reparametrizations:

(i) Unit speed curve of steepest descent with $\nabla f(\mathbf{x}(0)) \ne \mathbf{0}$:

$$\mathbf{y}'(s) = -\frac{\nabla f(\mathbf{y}(s))}{\|\nabla f(\mathbf{y}(s))\|_2}, \quad \mathbf{y}(0) = \mathbf{x}_0$$

with arc length S between $\mathbf{y}(s_0)$ and $\mathbf{y}(s_1)$:

$$S = \int_{s_0}^{s_1} \|\mathbf{y}'(s)\|_2 \, ds = \int_{s_0}^{s_1} 1 \, ds = s_1 - s_0.$$

(ii) Curve of steepest descent according to objective function values with $\nabla f(\mathbf{x}(0)) \ne \mathbf{0}$:

$$\mathbf{v}'(\rho) = -\frac{\nabla f(\mathbf{v}(\rho))}{\|\nabla f(\mathbf{v}(\rho))\|_2^2}, \quad \mathbf{v}(0) = \mathbf{x}_0$$

with

$$f(\mathbf{v}(\rho_1)) - f(\mathbf{v}(\rho_0)) = \int_{\rho_0}^{\rho_1} \frac{d}{d\rho} f(\mathbf{v}(\rho)) d\rho = \int_{\rho_0}^{\rho_1} (-1) d\rho = \rho_0 - \rho_1.$$

The application of reparametrizations to nonlinear programming is discussed in [SchäWar90].

A characteristic of a curve is its curvature, which measures the curve's local deviation from a straight line, for any curve point P. The curvature of a twice continuously differentiable curve

$$\mathbf{x} : [0, \infty) \to \mathbb{R}^2, \quad t \mapsto \mathbf{x}(t)$$

is defined by (see, e.g., [Tho78]):

$$\kappa : [0, \infty) \to \mathbb{R}, \quad t \mapsto \frac{\dot{x}_1(t)\ddot{x}_2(t) - \ddot{x}_1(t)\dot{x}_2(t)}{(\dot{x}_1(t)^2 + \dot{x}_2(t)^2)^{\frac{3}{2}}}.$$

We demonstrate the relevance of curvature by two examples.

Example 2.2. Consider the local minimization problem

$$\operatorname*{locmin}_{\mathbf{x}} \left\{ \frac{1}{2} \mathbf{x}^{\mathsf{T}} \underbrace{\begin{pmatrix} 1.5 & -0.5 \\ -0.5 & 1.5 \end{pmatrix}}_{\mathbf{M}} \mathbf{x} \right\}$$

with starting point $\mathbf{x}_0 = \begin{pmatrix} 2 \\ 0 \end{pmatrix}$.

The curve of steepest descent is given by

$$\dot{\mathbf{x}}(t) = -\begin{pmatrix} 1.5 & -0.5 \\ -0.5 & 1.5 \end{pmatrix} \mathbf{x}(t) \; (= -\mathbf{M}\mathbf{x}(t)), \quad \mathbf{x}(0) = \begin{pmatrix} 2 \\ 0 \end{pmatrix}$$

with unique solution (Fig. 2.1)

$$\mathbf{x} : [0, \infty) \to \mathbb{R}^2, \quad t \mapsto \begin{pmatrix} e^{-t} + e^{-2t} \\ e^{-t} - e^{-2t} \end{pmatrix}.$$

The curvature κ of this curve is shown in Fig. 2.2. Note that

$$0 < \kappa(t) < 1.5, \quad t \in [0, \infty).$$

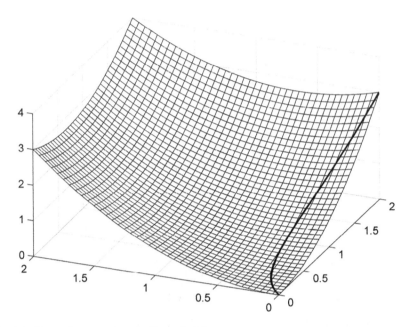

Fig. 2.1 Curve of steepest descent, Example 2.2

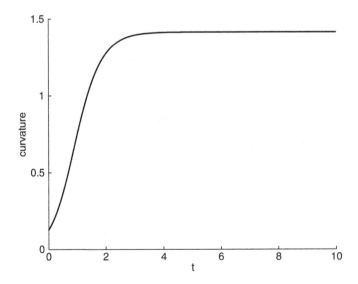

Fig. 2.2 Curvature, Example 2.2

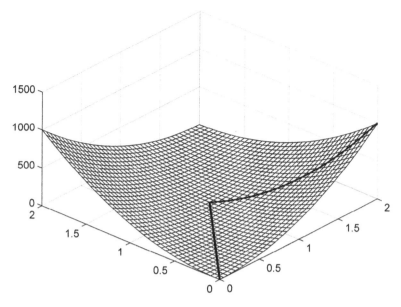

Fig. 2.3 Curve of steepest descent, Example 2.3

Example 2.3. Now, we investigate the local minimization problem

$$\operatorname*{locmin}_{\mathbf{x}} \left\{ \frac{1}{2}\mathbf{x}^{\top} \underbrace{\begin{pmatrix} 500.5 & -499.5 \\ -499.5 & 500.5 \end{pmatrix}}_{\mathbf{M}} \mathbf{x} \right\}$$

with starting point $\mathbf{x}_0 = \begin{pmatrix} 2 \\ 0 \end{pmatrix}$.

In this case, the curve of steepest descent is given by

$$\dot{\mathbf{x}}(t) = -\begin{pmatrix} 500.5 & -499.5 \\ -499.5 & 500.5 \end{pmatrix} \mathbf{x}(t) \ (= -\mathbf{M}\mathbf{x}(t)), \quad \mathbf{x}(0) = \begin{pmatrix} 2 \\ 0 \end{pmatrix}$$

with unique solution

$$\mathbf{x} : [0, \infty) \to \mathbb{R}^2, \quad t \mapsto \begin{pmatrix} e^{-t} + e^{-1000t} \\ e^{-t} - e^{-1000t} \end{pmatrix}.$$

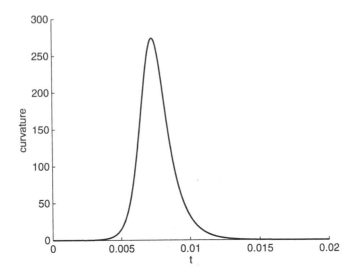

Fig. 2.4 Curvature, Example 2.3

The maximum curvature at $\hat{t} \approx 0.008$ with $\kappa(\hat{t}) \approx 275$ (Fig. 2.4) corresponds to the sharp bend of the curve in Fig. 2.3.

In general, it is not possible to compute a curve of steepest descent analytically. Therefore, we summarize numerical approximations of the curve of steepest descent in the next section. It turns out that the optimal choice of a numerical approximation of a curve of steepest descent depends essentially on the curvature of this curve.

2.2   Numerical Analysis

A self-evident approach for the numerical solution of the initial value problem

$$\dot{\mathbf{x}}(t) = -\nabla f(\mathbf{x}(t)), \quad \mathbf{x}(0) = \mathbf{x}_0,$$

is given by the Euler method. Assume that one has computed an approximation $\mathbf{x}_{app}(\bar{t})$ of $\mathbf{x}(\bar{t})$. The Euler method with step size $h > 0$ defines

$$\mathbf{x}_{app}(\bar{t} + h) = \mathbf{x}_{app}(\bar{t}) - h\nabla f(\mathbf{x}_{app}(\bar{t}))$$

as an approximation of $\mathbf{x}(\bar{t} + h)$. This method arises from the replacement of

$$\int_{\bar{t}}^{\bar{t}+h} \nabla f(\mathbf{x}(t))dt \quad \text{with} \quad h\nabla f(\mathbf{x}(\bar{t}))$$

in the integral form

$$\mathbf{x}(\bar{t} + h) = \mathbf{x}_{app}(\bar{t}) - \int_{\bar{t}}^{\bar{t}+h} \nabla f(\mathbf{x}(t))dt$$

of the initial value problem

$$\dot{\mathbf{x}}(t) = -\nabla f(\mathbf{x}(t)), \quad \mathbf{x}(\bar{t}) = \mathbf{x}_{app}(\bar{t}).$$

In terms of nonlinear programming, the Euler method is called method of steepest descent, where the step size h is chosen such that

$$f(\mathbf{x}_{app}(\bar{t} + h)) < f(\mathbf{x}_{app}(\bar{t})).$$

The Euler method is exact, if

$$\nabla f(\mathbf{x}(\bullet)) : [\bar{t}, \bar{t} + h] \to \mathbb{R}^n, \quad t \mapsto \nabla f(\mathbf{x}(t))$$

is a constant function. In this case, the curvature of the curve of steepest descent is identical to zero on $[\bar{t}, \bar{t} + h]$. Coming back to Example 2.3

$$\underset{\mathbf{x}}{\text{locmin}} \left\{ \frac{1}{2}\mathbf{x}^T \underbrace{\begin{pmatrix} 500.5 & -499.5 \\ -499.5 & 500.5 \end{pmatrix}}_{\mathbf{M}} \mathbf{x} \right\}$$

with starting point $\mathbf{x}_0 = \begin{pmatrix} 2 \\ 0 \end{pmatrix}$,

we show that the Euler method requires small step sizes in general (caused by the curvature of the curve of steepest descent). Using the Euler method with constant step size $h > 0$, one obtains with $\mathbf{I}_2 = \begin{pmatrix} 1 & 0 \\ 0 & 1 \end{pmatrix}$:

$$\begin{aligned}
\mathbf{x}_{app}(0 + ih) &= \mathbf{x}_{app}(0 + (i-1)h) - h\mathbf{M}\mathbf{x}_{app}(0 + (i-1)h) \\
&= (\mathbf{I}_2 - h\mathbf{M})\mathbf{x}_{app}(0 + (i-1)h) \\
&= (\mathbf{I}_2 - h\mathbf{M})^i \begin{pmatrix} 2 \\ 0 \end{pmatrix}.
\end{aligned}$$

While $\lim\limits_{t\to\infty} \mathbf{x}(t) = \begin{pmatrix} 0 \\ 0 \end{pmatrix}$, the sequence $\{\mathbf{x}_{\text{app}}(0 + ih)\}_{i \in \mathbb{N}}$ converges to $\begin{pmatrix} 0 \\ 0 \end{pmatrix}$ iff

$$|1 - h\lambda_1| < 1 \quad \text{and} \quad |1 - h\lambda_2| < 1$$

where $\lambda_1 = 1$ and $\lambda_2 = 1000$ are the eigenvalues of $\mathbf{M}$. Hence, the sequence $\{\mathbf{x}_{\text{app}}(0 + ih)\}_{i \in \mathbb{N}}$ converges to $\begin{pmatrix} 0 \\ 0 \end{pmatrix}$ iff $0 < h < 0.002$ (an analogous calculation leads to $0 < h < 1$ in Example 2.2).

In order to avoid these difficulties, one replaces

$$\int_{\bar{t}}^{\bar{t}+h} \nabla f(\mathbf{x}(t))dt \quad \text{with} \quad h\nabla f(\mathbf{x}(\bar{t} + h))$$

in the integral form

$$\mathbf{x}(\bar{t} + h) = \mathbf{x}_{\text{app}}(\bar{t}) - \int_{\bar{t}}^{\bar{t}+h} \nabla f(\mathbf{x}(t))dt$$

of the initial value problem

$$\dot{\mathbf{x}}(t) = -\nabla f(\mathbf{x}(t)), \quad \mathbf{x}(\bar{t}) = \mathbf{x}_{\text{app}}(\bar{t}),$$

which leads to the implicit Euler method. Applying this method to

$$\dot{\mathbf{x}}(t) = -\begin{pmatrix} 500.5 & -499.5 \\ -499.5 & 500.5 \end{pmatrix} \mathbf{x}(t) \ (= -\mathbf{M}\mathbf{x}(t)), \quad \mathbf{x}(0) = \begin{pmatrix} 2 \\ 0 \end{pmatrix}$$

yields

$$\mathbf{x}_{\text{app}}(0 + ih) = \mathbf{x}_{\text{app}}(0 + (i - 1)h) - h\mathbf{M}\mathbf{x}_{\text{app}}(0 + ih)$$

or alternatively,

$$\mathbf{x}_{\text{app}}(0 + ih) = (\mathbf{I}_2 + h\mathbf{M})^{-1}\mathbf{x}_{\text{app}}(0 + (i - 1)h)$$

$$= (\mathbf{I}_2 + h\mathbf{M})^{-i} \begin{pmatrix} 2 \\ 0 \end{pmatrix}.$$

Now, the sequence $\{\mathbf{x}_{\text{app}}(0 + ih)\}_{i \in \mathbb{N}}$ converges to $\begin{pmatrix} 0 \\ 0 \end{pmatrix}$ iff

$$|1 + h\lambda_1| > 1 \quad \text{and} \quad |1 + h\lambda_2| > 1$$

with $\lambda_1 = 1$ and $\lambda_2 = 1,000$ which means that there is no restriction on the step size any longer. Unfortunately, the implicit Euler method requires us to solve

$$\mathbf{x}_{\text{app}}(\bar{t} + h) = \mathbf{x}_{\text{app}}(\bar{t}) - h\nabla f(\mathbf{x}_{\text{app}}(\bar{t} + h))$$

or equivalently

$$\mathbf{x}_{\text{app}}(\bar{t} + h) + h\nabla f(\mathbf{x}_{\text{app}}(\bar{t} + h)) - \mathbf{x}_{\text{app}}(\bar{t}) = \mathbf{0}$$

which is a system of nonlinear equations in general. Considering the function

$$\mathbf{F} : \mathbb{R}^n \to \mathbb{R}^n, \quad \mathbf{z} \mapsto \mathbf{z} + h\nabla f(\mathbf{z}) - \mathbf{x}_{\text{app}}(\bar{t}),$$

the linearization of $\mathbf{F}$ at $\mathbf{x}_{\text{app}}(\bar{t})$ is given by

$$\mathbf{LF} : \mathbb{R}^n \to \mathbb{R}^n, \quad \mathbf{z} \mapsto h\nabla f(\mathbf{x}_{\text{app}}(\bar{t})) + \left(\mathbf{I}_n + h\nabla^2 f(\mathbf{x}_{\text{app}}(\bar{t}))\right)(\mathbf{z} - \mathbf{x}_{\text{app}}(\bar{t})),$$

where $\mathbf{I}_n$ denotes the n-dimensional identity matrix and $\nabla^2 f : \mathbb{R}^n \to \mathbb{R}^{n,n}$ is the Hessian of f. The equation

$$\mathbf{x}_{\text{app}}(\bar{t} + h) + h\nabla f(\mathbf{x}_{\text{app}}(\bar{t} + h)) - \mathbf{x}_{\text{app}}(\bar{t}) = \mathbf{0}$$

is equivalent to

$$\mathbf{F}(\mathbf{x}_{\text{app}}(\bar{t} + h)) = \mathbf{0}.$$

Replacing $\mathbf{F}$ by $\mathbf{LF}$ leads to

$$h\nabla f(\mathbf{x}_{\text{app}}(\bar{t})) + \left(\mathbf{I}_n + h\nabla^2 f(\mathbf{x}_{\text{app}}(\bar{t}))\right)(\mathbf{x}_{\text{app}}(\bar{t} + h) - \mathbf{x}_{\text{app}}(\bar{t})) = \mathbf{0}$$

or equivalently

$$\mathbf{x}_{\text{app}}(\bar{t} + h) = \mathbf{x}_{\text{app}}(\bar{t}) - \left(\frac{1}{h}\mathbf{I}_n + \nabla^2 f(\mathbf{x}_{\text{app}}(\bar{t}))\right)^{-1} \nabla f(\mathbf{x}_{\text{app}}(\bar{t}))$$

for suitable $h > 0$ (small enough such that $\left(\frac{1}{h}\mathbf{I}_n + \nabla^2 f(\mathbf{x}_{\text{app}}(\bar{t}))\right)$ is positive definite). This method is called semi-implicit Euler method and is closely related to the trust region method in nonlinear programming. In 1944, the addition of a multiple of the identity matrix to the Hessian as a stabilization procedure in the context of the solution of nonlinear least-squares problems appears to be published (in [Lev44]) for the first time by K. Levenberg (see [Conn.etal00] for a detailed description of trust region methods and their history). Furthermore,

$$\lim_{h \to \infty} -\left(\frac{1}{h}\mathbf{I}_n + \nabla^2 f(\mathbf{x}_{\text{app}}(\bar{t}))\right)^{-1} \nabla f(\mathbf{x}_{\text{app}}(\bar{t})) = -\left(\nabla^2 f(\mathbf{x}_{\text{app}}(\bar{t}))\right)^{-1} \nabla f(\mathbf{x}_{\text{app}}(\bar{t}))$$

for positive definite Hessian $\nabla^2 f(\mathbf{x}_{\text{app}}(\bar{t}))$. This is the Newtonian search direction.

Chapter 3
Unconstrained Global Optimization

3.1 A Randomized Curve of Steepest Descent

After a short discussion of local optimization, we investigate the following unconstrained global minimization problem:

$$\text{globmin}_{\mathbf{x}}\{f(\mathbf{x})\}, \quad f : \mathbb{R}^n \to \mathbb{R}, \ n \in \mathbb{N}, \quad f \in C^2(\mathbb{R}^n, \mathbb{R}).$$

In other words, we have to compute a point $\mathbf{x}_{gl} \in \mathbb{R}^n$ such that

$$f(\mathbf{x}) \geq f(\mathbf{x}_{gl}) \quad \text{for all } \mathbf{x} \in \mathbb{R}^n.$$

It is not reasonable to develop methods for solving unconstrained global minimization problems neglecting the existence of powerful methods solving unconstrained local minimization problems. Therefore, we are only interested in computing a suitable starting point close enough to a global minimum point in order to apply a local minimizing procedure.

Let the objective function $f : \mathbb{R}^n \to \mathbb{R}$ be such that

$$\int_{\mathbb{R}^n} \exp\left(-\frac{2f(\mathbf{x})}{\epsilon^2}\right) d\mathbf{x} < \infty$$

for some $\epsilon > 0$. In this case, the function

$$\lambda_f : \mathbb{R}^n \to \mathbb{R}, \quad \mathbf{x} \mapsto \frac{\exp\left(-\frac{2f(\mathbf{x})}{\epsilon^2}\right)}{\int_{\mathbb{R}^n} \exp\left(-\frac{2f(\mathbf{x})}{\epsilon^2}\right) d\mathbf{x}}$$

S. Schäffler, *Global Optimization: A Stochastic Approach*, Springer Series in Operations Research and Financial Engineering, DOI 10.1007/978-1-4614-3927-1_3, © Springer Science+Business Media New York 2012

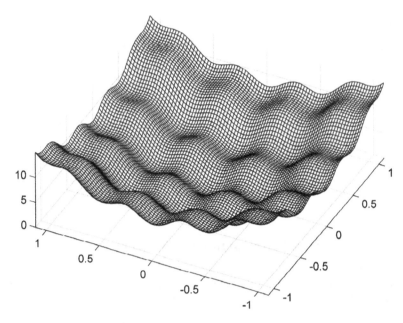

Fig. 3.1 Objective function f

can be interpreted as a Lebesgue density function of an n-dimensional real-valued random variable $\mathbf{X}_f$. We illustrate the main relationship between f and λ_f in the following example.

Example 3.1. The objective function

$$f : \mathbb{R}^2 \to \mathbb{R}, \quad (x, y)^\top \mapsto 6x^2 + 6y^2 - \cos(12x) - \cos(12y) + 2$$

has 25 isolated minimum points with the unique global minimum point at the origin (Figs. 3.1 and 3.2).

Figure 3.3 shows the corresponding Lebesgue density function λ_f.

The smaller the function value of f, the larger the function value of λ_f and finally the likelihood of $\mathbf{X}_f$. If we would be able to generate realizations of the random variable $\mathbf{X}_f$, we would be able to generate suitable starting points for local minimization procedures to compute a global minimum point of f. Unfortunately, there is no easy way to compute realizations of $\mathbf{X}_f$ in general. Therefore, we will use a dynamical system representing a suitable randomization of the curve of steepest descent in order to compute realizations of $\mathbf{X}_f$. For that, we have to introduce n-dimensional Brownian Motion.

Let

$$\Omega := C^0([0, \infty), \mathbb{R}^n)$$

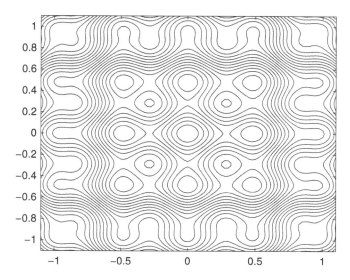

Fig. 3.2 Contour lines of f

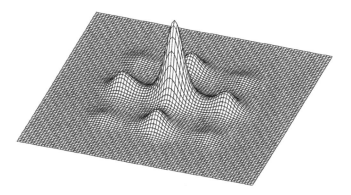

Fig. 3.3 Lebesgue density function λ_f

be the set of all continuous functions $\omega : [0, \infty) \to \mathbb{R}^n$ (for $t = 0$, continuity from the right). We define a metric d_Ω on Ω by

$$d_\Omega : \Omega \times \Omega \to \mathbb{R}, \quad (\omega_1, \omega_2) \mapsto \sum_{m=1}^{\infty} \frac{1}{2^m} \min \left\{ \max_{0 \leq t \leq m} \|\omega_1(t) - \omega_2(t)\|_2, 1 \right\}.$$

With $\mathcal{B}(\Omega)$, we denote the smallest σ-field containing all open subsets of Ω with respect to the topology defined by the metric d_Ω. Using the Borel σ-field $\mathcal{B}(\mathbb{R}^n)$ which is defined by all open subsets of $\mathbb{R}^n$ in the topology given by the Euclidean norm, we obtain $\mathcal{B}(\Omega)$-$\mathcal{B}(\mathbb{R}^n)$-measurable functions:

$$\mathbf{B}_t : \Omega \to \mathbb{R}^n, \quad \omega \mapsto \omega(t), \quad t \in [0, \infty).$$

There is a uniquely defined probability measure W on $\mathcal{B}(\Omega)$ (the so-called Wiener measure) fulfilling the following conditions, where $\mathcal{N}(\mathbf{e}, \boldsymbol{\Sigma})$ denotes the n-dimensional Gaussian distribution with expectation $\mathbf{e}$ and covariance matrix $\boldsymbol{\Sigma}$ (see, e.g., [KarShr08]):

(i) $W(\{\omega \in \Omega; \mathbf{B}_0(\omega) = \mathbf{0}\}) = 1$.
(ii) For all $0 \leq t_0 < t_1 < \ldots < t_k, k \in \mathbb{N}$, the random variables

$$\mathbf{B}_{t_0}, \mathbf{B}_{t_1} - \mathbf{B}_{t_0}, \ldots, \mathbf{B}_{t_k} - \mathbf{B}_{t_{k-1}}$$

are stochastically independent.
(iii) For every $0 \leq s < t$, the random variable $\mathbf{B}_t - \mathbf{B}_s$ is $\mathcal{N}(\mathbf{0}, (t - s)\mathbf{I}_n)$ Gaussian distributed.

The stochastic process $\{\mathbf{B}_t\}_{t \in [0,\infty)}$ is called n-dimensional Brownian Motion.
Based on the Wiener space, we will now investigate a stochastic process $\{\mathbf{X}_t\}_{t \in [0,\infty)}$,

$$\mathbf{X}_t : \Omega \to \mathbb{R}^n, \quad t \in [0, \infty),$$

given by

$$\mathbf{X}_t(\omega) = \mathbf{x}_0 - \int_0^t \nabla f(\mathbf{X}_\tau(\omega)) d\tau + \epsilon \left(\mathbf{B}_t(\omega) - \mathbf{B}_0(\omega)\right), \quad \omega \in \Omega,$$

where $\mathbf{x}_0 \in \mathbb{R}^n$ and $\epsilon > 0$ are fixed. This process can be interpreted as a randomized curve of steepest descent, where the curve of steepest descent in integral form is given by

$$\mathbf{x}(t) = \mathbf{x}_0 - \int_0^t \nabla f(\mathbf{x}(\tau)) d\tau, \quad t \in [0, \infty),$$

as noted above. The balance between steepest descent represented by

$$\mathbf{x}_0 - \int_0^t \nabla f(\mathbf{X}_\tau(\omega)) d\tau$$

and a purely random search using Gaussian distributed random variables

$$\mathbf{B}_t(\omega) - \mathbf{B}_0(\omega)$$

is controlled by the parameter ϵ.

In Sect. 2.1, we investigated the curve of steepest descent under weak assumptions on the objective function. In order to analyze the randomized curve of steepest descent, it is necessary again to formulate weak assumptions on the objective function:

Assumption 3.2 *There exists a real number $\epsilon > 0$ such that*

$$\mathbf{x}^\top \nabla f(\mathbf{x}) \geq \frac{1 + n\epsilon^2}{2} \max\{1, \|\nabla f(\mathbf{x})\|_2\}$$

for all $\mathbf{x} \in \{\mathbf{z} \in \mathbb{R}^n; \|\mathbf{z}\|_2 > \rho\}$ for some $\rho \in \mathbb{R}$, $\rho > 0$.

This assumption describes a special behavior of f outside a ball with radius ρ, for which only the existence is postulated. Starting at the origin, the objective function f has to increase sufficiently fast along each straight line outside the mentioned ball. Therefore, each function $f \in C^2(\mathbb{R}^n, \mathbb{R})$ fulfilling Assumption 3.2 has at least one global minimum point $\mathbf{x}_{gl}$ within the ball $\{\mathbf{z} \in \mathbb{R}^n; \|\mathbf{z}\|_2 \leq \rho\}$. The converse is not true as the $\sin$-function shows. The fact that only the existence of some possibly very large $\rho > 0$ is postulated ensures the weakness of this assumption.

In the theory of partial differential equations (see, e.g., [RenRog04]), $\mathbf{g} : \mathbb{R}^n \to \mathbb{R}^n$ is called a *coercive function* if

$$\lim_{\|\mathbf{x}\|_2 \to \infty} \frac{\mathbf{x}^\top \mathbf{g}(\mathbf{x})}{\|\mathbf{x}\|_2} = \infty.$$

The following alternative formulation of Assumption 3.2:

There exists a real number $\epsilon > 0$ such that

$$\frac{\mathbf{x}^\top \nabla f(\mathbf{x})}{\|\mathbf{x}\|_2} \geq \frac{1 + n\epsilon^2}{2} \max\left\{ \frac{1}{\|\mathbf{x}\|_2}, \frac{\|\nabla f(\mathbf{x})\|_2}{\|\mathbf{x}\|_2} \right\}$$

for all $\mathbf{x} \in \{\mathbf{z} \in \mathbb{R}^n; \|\mathbf{z}\|_2 > \rho\}$ for some $\rho \in \mathbb{R}$, $\rho > 0$

shows the relation of these two conditions on $\mathbf{g}$ and ∇f.

If Assumption 3.2 is not fulfilled, one can try to use an auxiliary objective function $\bar{f}$ of the following type:

$$\bar{f} : \mathbb{R}^n \to \mathbb{R}, \quad \mathbf{x} \mapsto f(\mathbf{x}) + \left(P\left(\|\mathbf{x}\|_2^2 - c \right) \right)^m, \ m \in \mathbb{N}, \ m \geq 3, \ c \in \mathbb{R}, \ c > 0,$$

where

$$P : \mathbb{R} \to \mathbb{R}, \quad x \mapsto \begin{cases} x & \text{for} \quad x > 0 \\ 0 & \text{for} \quad x \leq 0 \end{cases}.$$

For $\bar{f}$, we obtain:

- $\bar{f} \in C^2(\mathbb{R}^n, \mathbb{R})$.
- $\bar{f}(x) = f(x)$ for all $x \in \left\{ \mathbf{z} \in \mathbb{R}^n; \|\mathbf{z}\|_2^2 \leq c \right\}$.
- $\bar{f}(x) > f(x)$ for all $x \in \left\{ \mathbf{z} \in \mathbb{R}^n; \|\mathbf{z}\|_2^2 > c \right\}$.

The practicability of $\bar{f}$ instead of f is shown by

$$\mathbf{x}^T \nabla \bar{f}(\mathbf{x}) = \mathbf{x}^T \nabla f(\mathbf{x}) + 2m \left(P \left(\|\mathbf{x}\|_2^2 - c \right) \right)^{m-1} \|\mathbf{x}\|_2^2.$$

In the following theorem, we study properties of the randomized curve of steepest descent.

Theorem 3.3 *Consider*

$$f : \mathbb{R}^n \to \mathbb{R},\ n \in \mathbb{N},\quad f \in C^2(\mathbb{R}^n, \mathbb{R})$$

and let the following Assumption 3.2 be fulfilled:

There exists a real number $\epsilon > 0$ such that

$$\mathbf{x}^T \nabla f(\mathbf{x}) \geq \frac{1 + n\epsilon^2}{2} \max\{1, \|\nabla f(\mathbf{x})\|_2\}$$

for all $\mathbf{x} \in \{\mathbf{z} \in \mathbb{R}^n;\ \|\mathbf{z}\|_2 > \rho\}$ for some $\rho \in \mathbb{R}$, $\rho > 0$,

then we obtain:

(i) *Using the Wiener space $(\Omega, \mathcal{B}(\Omega), W)$ and $\epsilon > 0$ from Assumption 3.2, the integral equation*

$$\mathbf{X}(t, \omega) = \mathbf{x}_0 - \int_0^t \nabla f(\mathbf{X}(\tau, \omega)) d\tau + \epsilon \left(\mathbf{B}_t(\omega) - \mathbf{B}_0(\omega) \right),\quad t \in [0, \infty),\ \omega \in \Omega$$

has a unique solution $\mathbf{X} : [0, \infty) \times \Omega \to \mathbb{R}^n$ for each $\mathbf{x}_0 \in \mathbb{R}^n$.

(ii) *For each $t \in [0, \infty)$, the mapping*

$$\mathbf{X}_t : \Omega \to \mathbb{R}^n,\quad \omega \mapsto \mathbf{X}(t, \omega)$$

is an n-dimensional random variable (therefore, $\mathcal{B}(\Omega) - \mathcal{B}(\mathbb{R}^n)$ measurable) and its probability distribution is given by a Lebesgue density function

$$p_t : \mathbb{R}^n \to \mathbb{R}$$

with

$$\lim_{t \to \infty} p_t(\mathbf{x}) = \frac{\exp\left(-\frac{2f(\mathbf{x})}{\epsilon^2} \right)}{\int_{\mathbb{R}^n} \exp\left(-\frac{2f(\mathbf{x})}{\epsilon^2} \right) d\mathbf{x}}\quad \text{for all}\ \ \mathbf{x} \in \mathbb{R}^n.$$

The pointwise convergence of the Lebesgue density functions p_t to

$$\lambda_f : \mathbb{R}^n \to \mathbb{R}, \quad \mathbf{x} \mapsto \frac{\exp\left(-\frac{2f(\mathbf{x})}{\epsilon^2}\right)}{\int\limits_{\mathbb{R}^n} \exp\left(-\frac{2f(\mathbf{x})}{\epsilon^2}\right) d\mathbf{x}}$$

can be interpreted as follows:

The randomized curve of steepest descent can be interpreted as a machine learning scheme for the computation of pseudorandom vectors according to the probability distribution given by λ_f using evaluations of ∇f (or higher derivatives) and using the computation of standard Gaussian distributed pseudorandom numbers as inputs. This will be done by the numerical approximation of a path of the randomized curve of steepest descent.

For lucidity of the proof of Theorem 3.3, we first prove two lemmata.

Lemma 3.4 *Let*

$$\mathbf{g} : \mathbb{R}^n \to \mathbb{R}^n$$

be a globally Lipschitz-continuous function with Lipschitz constant $L > 0$, and let

$$\mathbf{B} : [0, \infty) \to \mathbb{R}^n$$

be a continuous function; then the integral equation

$$\mathbf{x}(t) = \mathbf{x}_0 - \int_0^t \mathbf{g}(\mathbf{x}(\tau)) d\tau + \mathbf{B}(t), \quad t \in [0, \infty),$$

has a unique solution

$$\mathbf{x} : [0, \infty) \to \mathbb{R}^n$$

for each $\mathbf{x}_0 \in \mathbb{R}^n$.

Proof. Since $\mathbf{g}$ is globally Lipschitz-continuous with Lipschitz constant $L > 0$, we prove existence and uniqueness of a solution $\mathbf{x} : [0, \infty) \to \mathbb{R}^n$ of the considered integral equation using the fixed point theorem of Banach. Let $C^0([0, T], \mathbb{R}^n)$ be the set of all continuous functions $\mathbf{u} : [0, T] \to \mathbb{R}^n$, and

$$K : C^0([0, T], \mathbb{R}^n) \to C^0([0, T], \mathbb{R}^n),$$

$$K(\mathbf{u})(t) = \mathbf{x}_0 - \int_0^t \mathbf{g}(\mathbf{u}(\tau)) d\tau + \mathbf{B}(t), \quad t \in [0, T].$$

Obviously, each solution $\mathbf{x}_T$ of

$$\mathbf{z}(t) = \mathbf{x}_0 - \int_0^t \mathbf{g}(\mathbf{z}(\tau)) d\tau + \mathbf{B}(t), \quad t \in [0, T]$$

is a fixed point of K and vice versa. With

$$d : C^0([0, T], \mathbb{R}^n) \times C^0([0, T], \mathbb{R}^n) \to \mathbb{R}, \quad (\mathbf{u}, \mathbf{v}) \mapsto \max_{t \in [0, T]} \left(\|\mathbf{u}(t) - \mathbf{v}(t)\|_2 e^{-2Lt} \right),$$

$(C^0([0, T], \mathbb{R}^n), d)$ is a complete metric space as mentioned before.

Again, the calculation

$$\|K(\mathbf{u})(t) - K(\mathbf{v})(t)\|_2 e^{-2Lt} = \left\| \int_0^t (\mathbf{g}(\mathbf{v}(\tau)) - \mathbf{g}(\mathbf{u}(\tau))) d\tau \right\|_2 e^{-2Lt}$$

$$\leq \int_0^t \|\mathbf{g}(\mathbf{v}(\tau)) - \mathbf{g}(\mathbf{u}(\tau))\|_2 d\tau \cdot e^{-2Lt}$$

$$= \int_0^t \|\mathbf{g}(\mathbf{v}(\tau)) - \mathbf{g}(\mathbf{u}(\tau))\|_2 e^{-2L\tau} e^{2L\tau} d\tau \cdot e^{-2Lt}$$

$$\leq L \int_0^t \|\mathbf{v}(\tau) - \mathbf{u}(\tau)\|_2 e^{-2L\tau} e^{2L\tau} d\tau \cdot e^{-2Lt}$$

$$\leq L \cdot d(\mathbf{u}, \mathbf{v}) \int_0^t e^{2L\tau} d\tau \cdot e^{-2Lt}$$

$$= L \cdot d(\mathbf{u}, \mathbf{v}) \frac{1}{2L} \left(e^{2Lt} - 1 \right) e^{-2Lt}$$

$$\leq \frac{L}{2L} d(\mathbf{u}, \mathbf{v}) = \frac{1}{2} d(\mathbf{u}, \mathbf{v}), \quad t \in [0, T]$$

shows that

$$d(K(\mathbf{u}), K(\mathbf{v})) \leq \frac{1}{2} d(\mathbf{u}, \mathbf{v}),$$

and that the fixed point theorem of Banach is applicable. Consequently, we have found a unique solution $\mathbf{x}_T : [0, T] \to \mathbb{R}^n$ of

$$\mathbf{z}(t) = \mathbf{x}_0 - \int_0^t \mathbf{g}(\mathbf{z}(\tau)) d\tau + \mathbf{B}(t), \quad t \in [0, T]$$

for all $T > 0$, and this gives us a unique solution $\mathbf{x} : [0, \infty) \to \mathbb{R}^n$ of

$$\mathbf{x}(t) = \mathbf{x}_0 - \int_0^t \mathbf{g}(\mathbf{x}(\tau)) d\tau + \mathbf{B}(t), \quad t \in [0, \infty),$$

q.e.d.

Lemma 3.5 *Let $f \in C^2(\mathbb{R}^n, \mathbb{R})$ and $\epsilon > 0$ such that Assumption 3.2 is fulfilled, then*

$$\int_{\mathbb{R}^n} \exp\left(-\frac{2f(\mathbf{x})}{\epsilon^2}\right) d\mathbf{x} < \infty.$$

Proof. For every $\mathbf{y} \in \mathbb{R}^n$, $\|\mathbf{y}\|_2 \neq \mathbf{0}$, we obtain for $\gamma > \rho$

$$\nabla f\left(\gamma \frac{\mathbf{y}}{\|\mathbf{y}\|_2}\right)^{\mathsf{T}} \gamma \frac{\mathbf{y}}{\|\mathbf{y}\|_2} \geq \frac{1 + n\epsilon^2}{2}$$

and therefore

$$\frac{d}{d\gamma} f\left(\gamma \frac{\mathbf{y}}{\|\mathbf{y}\|_2}\right) \geq \frac{1 + n\epsilon^2}{2\gamma}.$$

Integration over $[\rho, \xi]$, $\xi > \rho$, with respect to γ, leads to

$$f\left(\xi \frac{\mathbf{y}}{\|\mathbf{y}\|_2}\right) \geq \frac{1 + n\epsilon^2}{2} \ln(\xi) - \frac{1 + n\epsilon^2}{2} \ln(\rho) + f\left(\rho \frac{\mathbf{y}}{\|\mathbf{y}\|_2}\right).$$

Let

$$c := \min_{\|\mathbf{y}\|_2 \neq 0} \left\{ f\left(\rho \frac{\mathbf{y}}{\|\mathbf{y}\|_2}\right) \right\}.$$

For every $\mathbf{x} \in \{\mathbf{z} \in \mathbb{R}^n; \|\mathbf{z}\|_2 > \rho\}$, there exists a unique $\xi > \rho$ and a unique $\mathbf{y} \in \mathbb{R}^n$ with

$$\mathbf{x} = \xi \frac{\mathbf{y}}{\|\mathbf{y}\|_2}.$$

We obtain:

$$f(\mathbf{x}) \geq \frac{1 + n\epsilon^2}{2} \ln(\|\mathbf{x}\|_2) + c - \frac{1 + n\epsilon^2}{2} \ln(\rho)$$

for all $\mathbf{x} \in \{\mathbf{z} \in \mathbb{R}^n; \|\mathbf{z}\|_2 > \rho\}$. This inequality is equivalent to

$$\exp\left(-\frac{2f(\mathbf{x})}{\epsilon^2}\right) \leq c_1 \|\mathbf{x}\|_2^{-n-\frac{1}{\epsilon^2}}$$

with

$$c_1 = \exp\left(\frac{1 + n\epsilon^2}{\epsilon^2} \ln(\rho) - \frac{2c}{\epsilon^2}\right).$$

Integration leads to

$$
\int_{\mathbb{R}^n} \exp\left(-\frac{2f(\mathbf{x})}{\epsilon^2}\right) d\mathbf{x} = \int_{\{z\in\mathbb{R}^n;\, \|z\|_2 \le \rho\}} \exp\left(-\frac{2f(\mathbf{x})}{\epsilon^2}\right) d\mathbf{x}
$$

$$
+ \int_{\{z\in\mathbb{R}^n;\, \|z\|_2 > \rho\}} \exp\left(-\frac{2f(\mathbf{x})}{\epsilon^2}\right) d\mathbf{x}
$$

$$
\le \int_{\{z\in\mathbb{R}^n;\, \|z\|_2 \le \rho\}} \exp\left(-\frac{2f(\mathbf{x})}{\epsilon^2}\right) d\mathbf{x}
$$

$$
+ \int_{\{z\in\mathbb{R}^n;\, \|z\|_2 > \rho\}} c_1 \|\mathbf{x}\|_2^{-n-\frac{1}{\epsilon^2}} d\mathbf{x}
$$

$$
< \infty
$$

using n-dimensional polar coordinates for the computation of

$$
\int_{\{z\in\mathbb{R}^n;\, \|z\|_2 > \rho\}} c_1 \|\mathbf{x}\|_2^{-n-\frac{1}{\epsilon^2}} d\mathbf{x}.
$$

q.e.d.

Proof of Theorem 3.3. Using

$$
\mathbf{g} : \mathbb{R}^n \to \mathbb{R}^n, \quad \mathbf{x} \mapsto
\begin{cases}
\nabla f(\mathbf{x}) & \text{if} \quad \|\mathbf{x} - \mathbf{x}_0\|_2 \le r \\[2mm]
\nabla f\left(\mathbf{x}_0 + \frac{r(\mathbf{x}-\mathbf{x}_0)}{\|\mathbf{x}-\mathbf{x}_0\|_2}\right) & \text{if} \quad \|\mathbf{x} - \mathbf{x}_0\|_2 > r
\end{cases}
, \quad r > 0,
$$

we consider the integral equation

$$
\mathbf{Z}(t,\omega) = \mathbf{x}_0 - \int_0^t \mathbf{g}\left(\mathbf{Z}(\tau,\omega)\right) d\tau + \epsilon\left(\mathbf{B}_t(\omega) - \mathbf{B}_0(\omega)\right), \quad t \in [0,\infty), \quad \omega \in \Omega.
$$

Since $\mathbf{g}$ is globally Lipschitz-continuous with Lipschitz constant $L > 0$, and since each path of a Brownian Motion is continuous, the application of Lemma 3.4 for each $\omega \in \Omega$ shows the existence and uniqueness of a solution

$$
\mathbf{Z} : [0,\infty) \times \Omega \to \mathbb{R}^n
$$

of the above integral equation.

Now, we have to consider the connection between the function $\mathbf{Z}$ and the integral equation

$$\mathbf{X}(t,\omega) = \mathbf{x}_0 - \int_0^t \nabla f(\mathbf{X}(\tau,\omega))d\tau + \epsilon\left(\mathbf{B}_t(\omega) - \mathbf{B}_0(\omega)\right), \quad t \in [0,\infty), \quad \omega \in \Omega.$$

Therefore, we introduce the functions

$$s_r : \Omega \to \mathbb{R} \cup \{\infty\},$$

$$\omega \mapsto \begin{cases} \inf\{t \geq 0; \|\mathbf{Z}(t,\omega) - \mathbf{x}_0\|_2 \geq r\} & \text{if} \quad \{t \geq 0; \|\mathbf{Z}(t,\omega) - \mathbf{x}_0\|_2 \geq r\} \neq \emptyset \\ \infty & \text{if} \quad \{t \geq 0; \|\mathbf{Z}(t,\omega) - \mathbf{x}_0\|_2 \geq r\} = \emptyset \end{cases}$$

for every $r > 0$. Using s_r, it is obvious that the set of functions

$$\left\{\mathbf{Z}_{|s_r} : [0, s_r(\omega)) \to \mathbb{R}^n, t \mapsto \mathbf{Z}(t,\omega); \quad \omega \in \Omega\right\}$$

defines the unique solution of

$$\mathbf{X}(t,\omega) = \mathbf{x}_0 - \int_0^t \nabla f(\mathbf{X}(\tau,\omega))d\tau + \epsilon\left(\mathbf{B}_t(\omega) - \mathbf{B}_0(\omega)\right), \quad t \in [0, s_r(\omega)), \quad \omega \in \Omega.$$

Hence, we have to show that

$$\lim_{r \to \infty} s_r(\omega) = \infty \quad \text{for all } \omega \in \Omega.$$

For every $\omega \in \Omega$, we obtain a monotonically increasing function

$$s_\omega : [0,\infty) \to \mathbb{R} \cup \{\infty\}, \quad r \mapsto s_r(\omega).$$

Now, we assume the existence of $\hat{\omega} \in \Omega$ such that

$$\lim_{r \to \infty} s_{\hat{\omega}}(r) = \lim_{r \to \infty} s_r(\hat{\omega}) = s < \infty.$$

In order to lead this assumption to a contradiction, we consider the function

$$k : (0, s) \to \mathbb{R}, \quad t \mapsto \frac{d}{dt}\left(\frac{1}{2}\|\mathbf{Z}(t,\hat{\omega}) - \epsilon\left(\mathbf{B}_t(\hat{\omega}) - \mathbf{B}_0(\hat{\omega})\right) + \epsilon\left(\mathbf{B}_s(\hat{\omega}) - \mathbf{B}_0(\hat{\omega})\right)\|_2^2\right).$$

Choosing $\bar{t} \in [0, s)$ such that

- $\|\mathbf{Z}(\bar{t},\hat{\omega})\|_2 > \rho$ (ρ from Assumption 3.2),
- $\|\epsilon\left(\mathbf{B}_s(\hat{\omega}) - \mathbf{B}_t(\hat{\omega})\right)\|_2 < \frac{1+n\epsilon^2}{4}$ for all $t \in [\bar{t}, s)$,

we obtain for all $t \in [\bar{t}, s)$ with $\|\mathbf{Z}(t, \hat{\omega})\|_2 > \rho$

$$
\begin{aligned}
k(t) &= -\left(\mathbf{Z}(t, \hat{\omega}) + \epsilon \left(\mathbf{B}_s(\hat{\omega}) - \mathbf{B}_t(\hat{\omega})\right)\right)^\top \nabla f(\mathbf{Z}(t, \hat{\omega})) \\
&= -\mathbf{Z}(t, \hat{\omega})^\top \nabla f(\mathbf{Z}(t, \hat{\omega})) - \epsilon \left(\mathbf{B}_s(\hat{\omega}) - \mathbf{B}_t(\hat{\omega})\right)^\top \nabla f(\mathbf{Z}(t, \hat{\omega})) \\
&\leq -\frac{1 + n\epsilon^2}{2} \max\{1, \|\nabla f(\mathbf{Z}(t, \hat{\omega}))\|_2\} + \frac{1 + n\epsilon^2}{4} \|\nabla f(\mathbf{Z}(t, \hat{\omega}))\|_2 \\
&= -\frac{1 + n\epsilon^2}{4} \left(\max\{2, 2\|\nabla f(\mathbf{Z}(t, \hat{\omega}))\|_2\} - \|\nabla f(\mathbf{Z}(t, \hat{\omega}))\|_2\right) \\
&= -\frac{1 + n\epsilon^2}{4} \max\{2 - \|\nabla f(\mathbf{Z}(t, \hat{\omega}))\|_2, \|\nabla f(\mathbf{Z}(t, \hat{\omega}))\|_2\} \\
&\leq -\frac{1 + n\epsilon^2}{4} < 0.
\end{aligned}
$$

Consequently, for all $t \in [\bar{t}, s)$ holds:

$$
\begin{aligned}
\|\mathbf{Z}(t, \hat{\omega})\|_2 &= \|\mathbf{Z}(t, \hat{\omega}) + \epsilon \left(\mathbf{B}_s(\hat{\omega}) - \mathbf{B}_t(\hat{\omega})\right) - \epsilon \left(\mathbf{B}_s(\hat{\omega}) - \mathbf{B}_t(\hat{\omega})\right)\|_2 \\
&\leq \|\mathbf{Z}(t, \hat{\omega}) + \epsilon \left(\mathbf{B}_s(\hat{\omega}) - \mathbf{B}_t(\hat{\omega})\right)\|_2 + \|\epsilon \left(\mathbf{B}_s(\hat{\omega}) - \mathbf{B}_t(\hat{\omega})\right)\|_2 \\
&= \|\mathbf{Z}(t, \hat{\omega}) - \epsilon \left(\mathbf{B}_t(\hat{\omega}) - \mathbf{B}_0(\hat{\omega})\right) + \epsilon \left(\mathbf{B}_s(\hat{\omega}) - \mathbf{B}_0(\hat{\omega})\right)\|_2 \\
&\quad + \|\epsilon \left(\mathbf{B}_s(\hat{\omega}) - \mathbf{B}_t(\hat{\omega})\right)\|_2 \\
&\leq \|\mathbf{Z}(\bar{t}, \hat{\omega})\|_2 + \max_{\bar{t} \leq t \leq s} \{\|\epsilon \left(\mathbf{B}_s(\hat{\omega}) - \mathbf{B}_0(\hat{\omega})\right) - \epsilon \left(\mathbf{B}_t(\hat{\omega}) - \mathbf{B}_0(\hat{\omega})\right)\|_2\} \\
&\quad + \frac{1 + n\epsilon^2}{4}.
\end{aligned}
$$

This is a contradiction to

$$
\lim_{r \to \infty} \|\mathbf{Z}(s_r(\hat{\omega}), \hat{\omega}) - \mathbf{x}_0\|_2 = \infty
$$

and the proof of the first part of the theorem is finished.

Now, we choose $t \in (0, \infty)$, $m \in \mathbb{N}$, and $t_j := j\frac{t}{m}$, $j = 0, \ldots, m$. Since

$$
\mathbf{X}_t : \Omega \to \mathbb{R}^n, \quad \omega \mapsto \mathbf{X}_t(\omega)
$$

is the limit of a fixed point iteration

$$
\mathbf{X}_t(\omega) = \lim_{k \to \infty} \mathbf{X}_t^k(\omega)
$$

with

- $\mathbf{X}_t^0 : \Omega \to \mathbb{R}^n, \quad \omega \mapsto \mathbf{x}_0,$
- $\mathbf{X}_t^k : \Omega \to \mathbb{R}^n, \quad \omega \mapsto \mathbf{x}_0 - \int\limits_0^t \nabla f\left(\mathbf{X}_\tau^{k-1}(\omega)\right) d\tau + \epsilon \left(\mathbf{B}_t(\omega) - \mathbf{B}_0(\omega)\right),$

and since

$$\int_0^t \nabla f\left(\mathbf{X}_\tau^{k-1}(\omega)\right) d\tau = \lim_{m\to\infty} \sum_{j=1}^m \nabla f\left(\mathbf{X}_{t_{j-1}}^{k-1}(\omega)\right)(t_j - t_{j-1}),$$

each function $\mathbf{X}_t$ is $\mathcal{B}(\Omega) - \mathcal{B}(\mathbb{R}^n)$ measurable. Existence and properties of the Lebesgue density functions p_t are proven by Lemma 3.5 in combination with the analysis of the Cauchy problem for parabolic equations (see, e.g. [Fried06], Chap. 6, Sect. 4). **q.e.d.**

From Theorem 3.3, we know that choosing any starting point $\mathbf{x}_0 \in \mathbb{R}^n$, the numerical computation of a path

$$\mathbf{X}_{\tilde{\omega}} : [0, \infty) \to \mathbb{R}^n, \quad t \mapsto \mathbf{X}_t(\tilde{\omega})$$

with

$$\mathbf{X}_t(\tilde{\omega}) = \mathbf{x}_0 - \int_0^t \nabla f(\mathbf{X}_\tau(\tilde{\omega}))d\tau + \epsilon\left(\mathbf{B}_t(\tilde{\omega}) - \mathbf{B}_0(\tilde{\omega})\right), \quad t \in [0, \infty)$$

leads to a realization of $\mathbf{X}_f$ with Lebesgue density function

$$\lambda_f : \mathbb{R}^n \to \mathbb{R}, \quad \mathbf{x} \mapsto \frac{\exp\left(-\frac{2f(\mathbf{x})}{\epsilon^2}\right)}{\int_{\mathbb{R}^n} \exp\left(-\frac{2f(\mathbf{x})}{\epsilon^2}\right) d\mathbf{x}}.$$

Furthermore, we obtain from the stability theory of stochastic differential equations (see [Has80], Chap. 3, Theorem 7.1):

Theorem 3.6 *Let*

$$f : \mathbb{R}^n \to \mathbb{R}, n \in \mathbb{N}, \quad f \in C^2(\mathbb{R}^n, \mathbb{R}),$$

and let Assumption 3.2 be fulfilled. Choose any $r > 0$ and let $\mathbf{x}_{gl}$ be any global minimum point of f. Using

- *The Wiener space $(\Omega, \mathcal{B}(\Omega), W)$,*
- *Any $\epsilon > 0$ from Assumption 3.2,*
- *The integral equation*

$$\mathbf{X}(t, \omega) = \mathbf{x}_0 - \int_0^t \nabla f(\mathbf{X}(\tau, \omega))d\tau + \epsilon\left(\mathbf{B}_t(\omega) - \mathbf{B}_0(\omega)\right), \quad t \in [0, \infty), \quad \omega \in \Omega$$

with unique solution $\mathbf{X}$ according to Theorem 3.3,

- *The stopping time*

$$st \ : \ \Omega \to \mathbb{R} \cup \{\infty\},$$

$$\omega \mapsto \begin{cases} \inf\{t \ge 0; \ \|\mathbf{X}(t, \omega) - \mathbf{x}_{gl}\|_2 \le r\} & \text{if} \quad \{t \ge 0; \ \|\mathbf{X}(t, \omega) - \mathbf{x}_{gl}\|_2 \le r\} \ne \emptyset \\ \infty & \text{if} \quad \{t \ge 0; \ \|\mathbf{X}(t, \omega) - \mathbf{x}_{gl}\|_2 \le r\} = \emptyset \end{cases},$$

we obtain the following results:

(i) $W(\{\omega \in \Omega; \ st(\omega) < \infty\}) = 1.$
(ii) *For the expectation of st holds*

$$\mathbb{E}(st) < \infty.$$

Theorem 3.6 shows that almost all paths of the stochastic process $\{\mathbf{X}_t\}_{t \in [0,\infty)}$ generated by

$$\mathbf{X}_t(\omega) = \mathbf{x}_0 - \int_0^t \nabla f(\mathbf{X}_\tau(\omega)) d\tau + \epsilon \left(\mathbf{B}_t(\omega) - \mathbf{B}_0(\omega) \right), \quad t \in [0, \infty), \quad \omega \in \Omega$$

approximate each global minimum point of f with arbitrary accuracy in finite time.

3.2 Concepts of Numerical Analysis

As pointed out above, the solution of

$$\mathbf{X}_t(\tilde{\omega}) = \mathbf{x}_0 - \int_0^t \nabla f(\mathbf{X}_\tau(\tilde{\omega})) d\tau + \epsilon \left(\mathbf{B}_t(\tilde{\omega}) - \mathbf{B}_0(\tilde{\omega}) \right)$$

can be interpreted as a path of a randomized curve of steepest descent, where the curve of steepest descent in integral form is given by

$$\mathbf{x}(t) = \mathbf{x}_0 - \int_0^t \nabla f(\mathbf{x}(\tau)) d\tau, \quad t \in [0, \infty).$$

The balance between steepest descent represented by

$$\mathbf{x}_0 - \int_0^t \nabla f(\mathbf{X}_\tau(\tilde{\omega})) d\tau$$

and a purely random search using Gaussian distributed random variables represented by

$$\mathbf{B}_t(\tilde{\omega}) - \mathbf{B}_0(\tilde{\omega})$$

is controlled by the parameter ϵ.

The optimal choice of $\epsilon > 0$ (such that Assumption 3.2 is fulfilled) depends on the scaling of the objective function and has to be guided by the following considerations:

- If a long time is spent on the chosen path close to (local) minimum points of the objective function, then local minimization dominates and ϵ has to be increased.
- If minimum points of the objective function play no significant role along the path, then purely random search dominates and ϵ has to be decreased.

One can try to replace the parameter ϵ with a function

$$\epsilon : [0, \infty) \to \mathbb{R}, \quad t \mapsto \epsilon(t)$$

such that

1. $\epsilon(t) > 0$ for all $t \geq 0$,
2. $\lim_{t \to \infty} \epsilon(t) = 0$,

hoping that the Lebesgue density function

$$\lambda_f : \mathbb{R}^n \to \mathbb{R}, \quad \mathbf{x} \mapsto \frac{\exp\left(-\frac{2f(\mathbf{x})}{\epsilon^2}\right)}{\int_{\mathbb{R}^n} \exp\left(-\frac{2f(\mathbf{x})}{\epsilon^2}\right) d\mathbf{x}}$$

will converge pointwise to the point measure on all global minimum points (cooling strategy). But this strategy will only work if

- $\epsilon(0) > C$, where C is a constant depending on f.
- $\lim_{t \to \infty} \epsilon^2(t) \ln(t) > 0$.

If one of these conditions is not fulfilled, we find only local minimum points in general (as shown in [GemHwa86]).

Since almost all paths of the stochastic process $\{\mathbf{X}_t\}_{t \in [0,\infty)}$ are continuous but nowhere differentiable, classical strategies of step size control in numerical analysis are not applicable. Therefore, a step size control based on following principle is used:

Using a computed approximation $\mathbf{x}_{\text{app}}(\bar{t}, \tilde{\omega})$ of a path at $\bar{t}$, an approximation $\mathbf{x}_{\text{app}}(\bar{t} + h, \tilde{\omega})$ is computed twice, in one step with step size h and in two steps with step sizes $\frac{h}{2}$. If both approximations are close enough to each other, the step is accepted; otherwise, the approach must be repeated with $\frac{h}{2}$ instead of h. This step size control is based on the assumption that the exact solution is sufficiently close to the computed approximations, if these approximations are close enough to each other.

3.3 A Semi-implicit Euler Method

In this section, we investigate a semi-implicit Euler method for the numerical solution of

$$\mathbf{X}_t(\tilde{\omega}) = \mathbf{x}_0 - \int_0^t \nabla f(\mathbf{X}_\tau(\tilde{\omega})) d\tau + \epsilon \left(\mathbf{B}_t(\tilde{\omega}) - \mathbf{B}_0(\tilde{\omega}) \right), \quad t \in [0, \infty).$$

The implicit Euler method based on an approximation $\mathbf{x}_{\text{app}}(\bar{t}, \tilde{\omega})$ of $\mathbf{X}_{\bar{t}}(\tilde{\omega})$ leads to a system

$$\mathbf{x}_{\text{app}}(\bar{t} + h, \tilde{\omega}) = \mathbf{x}_{\text{app}}(\bar{t}, \tilde{\omega}) - h \nabla f(\mathbf{x}_{\text{app}}(\bar{t} + h, \tilde{\omega})) + \epsilon \left(\mathbf{B}_{\bar{t}+h}(\tilde{\omega}) - \mathbf{B}_{\bar{t}}(\tilde{\omega}) \right)$$

or equivalently

$$\mathbf{x}_{\text{app}}(\bar{t} + h, \tilde{\omega}) - \mathbf{x}_{\text{app}}(\bar{t}, \tilde{\omega}) + h \nabla f(\mathbf{x}_{\text{app}}(\bar{t} + h, \tilde{\omega})) - \epsilon \left(\mathbf{B}_{\bar{t}+h}(\tilde{\omega}) - \mathbf{B}_{\bar{t}}(\tilde{\omega}) \right) = \mathbf{0}$$

of nonlinear equations in general. We consider the linearization of

$$\mathbf{F} : \mathbb{R}^n \to \mathbb{R}^n, \quad \mathbf{z} \mapsto \mathbf{z} - \mathbf{x}_{\text{app}}(\bar{t}, \tilde{\omega}) + h \nabla f(\mathbf{z}) - \epsilon \left(\mathbf{B}_{\bar{t}+h}(\tilde{\omega}) - \mathbf{B}_{\bar{t}}(\tilde{\omega}) \right),$$

at $\mathbf{x}_{\text{app}}(\bar{t}, \tilde{\omega})$, which is given by

$$\mathbf{LF} : \mathbb{R}^n \to \mathbb{R}^n,$$

$$\mathbf{z} \mapsto h \nabla f(\mathbf{x}_{\text{app}}(\bar{t}, \tilde{\omega})) + \left(\mathbf{I}_n + h \nabla^2 f(\mathbf{x}_{\text{app}}(\bar{t}, \tilde{\omega})) \right) (\mathbf{z} - \mathbf{x}_{\text{app}}(\bar{t}, \tilde{\omega}))$$

$$- \epsilon \left(\mathbf{B}_{\bar{t}+h}(\tilde{\omega}) - \mathbf{B}_{\bar{t}}(\tilde{\omega}) \right).$$

Solving $\mathbf{LF} = \mathbf{0}$ instead of $\mathbf{F} = \mathbf{0}$ leads to

$$\mathbf{x}_{\text{app}}(\bar{t} + h, \tilde{\omega}) = \mathbf{x}_{\text{app}}(\bar{t}, \tilde{\omega})$$

$$- \left(\frac{1}{h} \mathbf{I}_n + \nabla^2 f(\mathbf{x}_{\text{app}}(\bar{t}, \tilde{\omega})) \right)^{-1} \left(\nabla f(\mathbf{x}_{\text{app}}(\bar{t}, \tilde{\omega})) - \frac{\epsilon}{h} \left(\mathbf{B}_{\bar{t}+h}(\tilde{\omega}) - \mathbf{B}_{\bar{t}}(\tilde{\omega}) \right) \right)$$

for small enough $h > 0$ (at least such that $\left(\frac{1}{h} \mathbf{I}_n + \nabla^2 f(\mathbf{x}_{\text{app}}(\bar{t}, \tilde{\omega})) \right)$ is positive definite).

Since $(\mathbf{B}_{\bar{t}+h} - \mathbf{B}_{\bar{t}})$ is a $\mathcal{N}(\mathbf{0}, h\mathbf{I}_n)$ Gaussian distributed random variable, the numerical evaluation of $\frac{\epsilon}{h} \left(\mathbf{B}_{\bar{t}+h}(\tilde{\omega}) - \mathbf{B}_{\bar{t}}(\tilde{\omega}) \right)$ can be done by algorithmic generation of n stochastically independent $\mathcal{N}(0, 1)$ Gaussian distributed pseudorandom numbers $p_1, \dots, p_n \in \mathbb{R}$. Hence, the vector $\frac{\epsilon}{h} \left(\mathbf{B}_{\bar{t}+h}(\tilde{\omega}) - \mathbf{B}_{\bar{t}}(\tilde{\omega}) \right)$ can be realized by

$$\frac{\epsilon}{h} \left(\mathbf{B}_{\bar{t}+h}(\tilde{\omega}) - \mathbf{B}_{\bar{t}}(\tilde{\omega}) \right) = \frac{\epsilon}{\sqrt{h}} \begin{pmatrix} p_1 \\ \vdots \\ p_n \end{pmatrix}.$$

Assume that we have computed $\mathbf{x}_{\text{app}}(h, \tilde{\omega})$ by a step with fixed step size h based on the starting point $\mathbf{x}_0 \in \mathbb{R}^n$ as described above:

$$\mathbf{x}_{\text{app}}(h, \tilde{\omega}) = \mathbf{x}_0 - \left(\frac{1}{h} \mathbf{I}_n + \nabla^2 f(\mathbf{x}_0) \right)^{-1} \left(\nabla f(\mathbf{x}_0) - \frac{\epsilon}{\sqrt{h}} \begin{pmatrix} p_1 \\ \vdots \\ p_n \end{pmatrix} \right).$$

Prima facie, the choice of $\tilde{\omega} \in \Omega$ is left to the computer by computing $p_1, \ldots, p_n \in \mathbb{R}$. But let us consider the set

$$\Omega_h := \{ \omega \in \Omega;\ \mathbf{X}_h(\omega) = \mathbf{X}_h(\tilde{\omega}) \}.$$

Apparently, the vector $\mathbf{x}_{\text{app}}(h, \tilde{\omega})$ is not only an approximation of $\mathbf{X}_h(\tilde{\omega})$ but also of $\mathbf{X}_h(\omega)$ for all $\omega \in \Omega_h$. Consequently, the computation of $p_1, \ldots, p_n \in \mathbb{R}$ does not cause a determination of $\tilde{\omega} \in \Omega$ but only a reduction of Ω to Ω_h.

Assume that we would like to compute $\mathbf{x}_{\text{app}}(2h, \tilde{\omega})$ by a step with fixed step size h based on the point $\mathbf{x}_{\text{app}}(h, \tilde{\omega})$. Therefore, we have to compute again n stochastically independent $\mathcal{N}(0, 1)$ Gaussian distributed pseudorandom numbers $q_1, \ldots, q_n \in \mathbb{R}$ and

$$\mathbf{x}_{\text{app}}(2h, \tilde{\omega}) = \mathbf{x}_{\text{app}}(h, \tilde{\omega})$$

$$- \left(\frac{1}{h} \mathbf{I}_n + \nabla^2 f(\mathbf{x}_{\text{app}}(h, \tilde{\omega})) \right)^{-1} \left(\nabla f(\mathbf{x}_{\text{app}}(h, \tilde{\omega})) - \frac{\epsilon}{\sqrt{h}} \begin{pmatrix} q_1 \\ \vdots \\ q_n \end{pmatrix} \right).$$

Since the random variables $(\mathbf{B}_{2h} - \mathbf{B}_h)$ and $(\mathbf{B}_h - \mathbf{B}_0)$ are stochastically independent, it is allowed to compute $q_1, \ldots, q_n$ independent of $p_1, \ldots, p_n$. The vector $\mathbf{x}_{\text{app}}(2h, \tilde{\omega})$ is a numerical approximation of $\mathbf{X}_{2h}(\omega)$ for all $\omega \in \Omega_{2h}$ with

$$\Omega_{2h} := \{ \omega \in \Omega;\ \mathbf{X}_h(\omega) = \mathbf{X}_h(\tilde{\omega}) \text{ and } \mathbf{X}_{2h}(\omega) = \mathbf{X}_{2h}(\tilde{\omega}) \}.$$

Since the function

$$\mathbf{X}_t(\tilde{\omega}) : [0, \infty) \to \mathbb{R}^n$$

is continuous but nowhere differentiable for almost all $\tilde{\omega} \in \Omega$, classical strategies of step size control in numerical analysis are not applicable. Therefore, we present a step size control based on the principle introduced in the last section. Beginning with a starting value h_{max} for h such that

$$\left(\frac{1}{h_{\text{max}}} \mathbf{I}_n + \nabla^2 f(\mathbf{x}_{\text{app}}(\bar{t}, \tilde{\omega})) \right)$$

is positive definite, we compute

$$\tilde{\mathbf{x}}\left(\bar{t} + \frac{h_{\max}}{2}\right) := \mathbf{x}_{\mathrm{app}}\left(\bar{t} + \frac{h_{\max}}{2}, \tilde{\omega}\right) = \mathbf{x}_{\mathrm{app}}(\bar{t}, \tilde{\omega})$$

$$- \left(\frac{1}{\frac{h_{\max}}{2}}\mathbf{I}_n + \nabla^2 f(\mathbf{x}_{\mathrm{app}}(\bar{t}, \tilde{\omega}))\right)^{-1} \left(\nabla f(\mathbf{x}_{\mathrm{app}}(\bar{t}, \tilde{\omega})) - \frac{\epsilon}{\sqrt{\frac{h_{\max}}{2}}}\begin{pmatrix} p_1 \\ \vdots \\ p_n \end{pmatrix}\right),$$

$$\tilde{\mathbf{x}}^1(\bar{t} + h_{\max}) := \mathbf{x}_{\mathrm{app}}^1(\bar{t} + h_{\max}, \tilde{\omega}) = \tilde{\mathbf{x}}\left(\bar{t} + \frac{h_{\max}}{2}\right)$$

$$- \left(\frac{1}{\frac{h_{\max}}{2}}\mathbf{I}_n + \nabla^2 f\left(\tilde{\mathbf{x}}\left(\bar{t} + \frac{h_{\max}}{2}\right)\right)\right)^{-1} \left(\nabla f\left(\tilde{\mathbf{x}}\left(\bar{t} + \frac{h_{\max}}{2}\right)\right) - \frac{\epsilon}{\sqrt{\frac{h_{\max}}{2}}}\begin{pmatrix} q_1 \\ \vdots \\ q_n \end{pmatrix}\right),$$

and

$$\tilde{\mathbf{x}}^2(\bar{t} + h_{\max}) := \mathbf{x}_{\mathrm{app}}^2(\bar{t} + h_{\max}, \tilde{\omega}) = \mathbf{x}_{\mathrm{app}}(\bar{t}, \tilde{\omega})$$

$$- \left(\frac{1}{h_{\max}}\mathbf{I}_n + \nabla^2 f(\tilde{\mathbf{x}}(\bar{t}))\right)^{-1} \left(\nabla f(\mathbf{x}_{\mathrm{app}}(\bar{t}, \tilde{\omega})) - \frac{\epsilon}{\sqrt{2h_{\max}}}\begin{pmatrix} p_1 + q_1 \\ \vdots \\ p_n + q_n \end{pmatrix}\right),$$

The vectors $\tilde{\mathbf{x}}^1(\bar{t} + h_{\max})$ and $\tilde{\mathbf{x}}^2(\bar{t} + h_{\max})$ represent two numerical approximations of $\mathbf{X}_{\bar{t}+h_{\max}}(\tilde{\omega})$. The approximation $\tilde{\mathbf{x}}^1(\bar{t} + h_{\max})$ is computed by two $\frac{h_{\max}}{2}$ steps based on $\mathbf{x}_{\mathrm{app}}(\bar{t}, \tilde{\omega})$ and using the random numbers $p_1, \ldots, p_n$ for the first step and $q_1, \ldots, q_n$ for the second step.

The approximation $\tilde{\mathbf{x}}^2(\bar{t} + h_{\max})$ is computed by one $h_{\max}$ step based on $\mathbf{x}_{\mathrm{app}}(\bar{t}, \tilde{\omega})$ and using the random numbers $p_1, \ldots, p_n$ and $q_1, \ldots, q_n$. This is necessary because we have to compute different approximations of the same path. Chosen any $\delta > 0$, we accept $\tilde{\mathbf{x}}^1(\bar{t} + h_{\max})$ as a numerical approximation of $\mathbf{X}_{\bar{t}+h_{\max}}(\tilde{\omega})$ if

$$\|\tilde{\mathbf{x}}^1(\bar{t} + h_{\max}) - \tilde{\mathbf{x}}^2(\bar{t} + h_{\max})\|_2 < \delta.$$

Otherwise, we have to repeat the step size procedure with $h = \frac{h_{\max}}{2}$ (Fig. 3.4).

The following algorithm describes a semi-implicit Euler method for the computation of

$$\mathbf{X}_\bullet(\tilde{\omega}) : [0, \infty) \to \mathbb{R}^n, \quad t \mapsto \mathbf{X}_t(\tilde{\omega}).$$

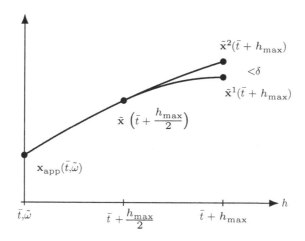

Fig. 3.4 Step size control

Step 0: (Initialization)

Choose $\mathbf{x}_0 \in \mathbb{R}^n$ and $\epsilon, \delta > 0$,
Choose maxit $\in \mathbb{N}$,
$j := 0$,
goto step 1.

In step 0, the starting point $\mathbf{x}_0$, the parameter ϵ according to
assumption 3.2, the parameter $\delta > 0$ according to the step size
control, and the maximal number of iterations have to be
determined by the user.

Step 1: (Derivatives)

$h := 1$,
compute $\nabla f(\mathbf{x}_j), \nabla^2 f(\mathbf{x}_j)$,
goto step 2.

The initial value h_{max} of the step size is chosen equal to 1.

Step 2: (Pseudorandom Numbers)

Compute $2n$ stochastically independent $\mathcal{N}(0, 1)$ Gaussian distributed
pseudorandom numbers $p_1, \ldots p_n, q_1, \ldots, q_n \in \mathbb{R}$,
goto step 3.

In this step, the choice of the path is determined successively
by the computer.

Step 3: (Cholesky Decomposition)

If $\left(\frac{1}{h}\mathbf{I}_n + \nabla^2 f\left(\mathbf{x}_j\right)\right) \in \mathbb{R}^{n,n}$ is positive definite, then

compute $\mathbf{L} \in \mathbb{R}^{n,n}$ such that:
$\mathbf{L}\mathbf{L}^\top = \left(\frac{1}{h}\mathbf{I}_n + \nabla^2 f\left(\mathbf{x}_j\right)\right)$ (Cholesky),
goto step 4.

else

$h := \frac{h}{2}$,
goto step 3.

Step 4: (Computation of $\mathbf{x}_{j+1}^2$ by one step with step size h)

Compute $\mathbf{x}_{j+1}^2$ by solving

$$\mathbf{L}\mathbf{L}^\top \mathbf{x}_{j+1}^2 = \left(\nabla f\left(\mathbf{x}_j\right) - \frac{\epsilon}{\sqrt{2h}}\begin{pmatrix} p_1 + q_1 \\ \vdots \\ p_n + q_n \end{pmatrix}\right),$$

$\mathbf{x}_{j+1}^2 := \mathbf{x}_j - \mathbf{x}_{j+1}^2$,
goto step 5.

$\mathbf{x}_{j+1}^2$ is computed by a step with starting point $\mathbf{x}_j$ using the step size h.

Step 5: (Cholesky Decomposition)

Compute $\mathbf{L} \in \mathbb{R}^{n,n}$ such that:
$\mathbf{L}\mathbf{L}^\top = \left(\frac{2}{h}\mathbf{I}_n + \nabla^2 f\left(\mathbf{x}_j\right)\right)$,
goto step 6.

The matrix $\left(\frac{2}{h}\mathbf{I}_n + \nabla^2 f\left(\mathbf{x}_j\right)\right)$ is obviously positive definite.

Step 6: (Computation of $\mathbf{x}_{\frac{h}{2}}$)

Compute $\mathbf{x}_{\frac{h}{2}}$ by solving

$$\mathbf{L}\mathbf{L}^\top \mathbf{x}_{\frac{h}{2}} = \left(\nabla f\left(\mathbf{x}_j\right) - \frac{\epsilon}{\sqrt{\frac{h}{2}}}\begin{pmatrix} p_1 \\ \vdots \\ p_n \end{pmatrix}\right),$$

$\mathbf{x}_{\frac{h}{2}} := \mathbf{x}_j - \mathbf{x}_{\frac{h}{2}}$,
goto step 7.

$\mathbf{x}_{\frac{h}{2}}$ is computed by a step with starting point $\mathbf{x}_j$ using the step size $\frac{h}{2}$.

Step 7: (Derivatives)

Compute $\nabla f(\mathbf{x}_{\frac{h}{2}})$, $\nabla^2 f(\mathbf{x}_{\frac{h}{2}})$,
goto step 8.

Step 8: (Cholesky Decomposition)

If $\left(\frac{2}{h}\mathbf{I}_n + \nabla^2 f\left(\mathbf{x}_{\frac{h}{2}}\right)\right) \in \mathbb{R}^{n,n}$ is positive definite, then

compute $\mathbf{L} \in \mathbb{R}^{n,n}$ such that:
$$\mathbf{L}\mathbf{L}^\top = \left(\tfrac{2}{h}\mathbf{I}_n + \nabla^2 f\left(\mathbf{x}_{\frac{h}{2}}\right) \right) \quad \text{(Cholesky)},$$
goto step 9.

else

$h := \frac{h}{2}$,
goto step 3.

Step 9: (Computation of $\mathbf{x}^1_{j+1}$ by two steps with step size $\frac{h}{2}$)

Compute $\mathbf{x}^1_{j+1}$ by solving

$$\mathbf{L}\mathbf{L}^\top \mathbf{x}^1_{j+1} = \left(\nabla f\left(\mathbf{x}_{\frac{h}{2}}\right) - \frac{\epsilon}{\sqrt{\frac{h}{2}}} \begin{pmatrix} q_1 \\ \vdots \\ q_n \end{pmatrix} \right),$$

$$\mathbf{x}^1_{j+1} := \mathbf{x}_{\frac{h}{2}} - \mathbf{x}^1_{j+1},$$

goto step 10.

$\mathbf{x}^1_{j+1}$ is computed by a step with starting point $\mathbf{x}_{\frac{h}{2}}$ using the step size $\frac{h}{2}$.

Step 10: (Acceptance condition)

If $\left\| \mathbf{x}^1_{j+1} - \mathbf{x}^2_{j+1} \right\|_2 < \delta$,

then

$\mathbf{x}_{j+1} := \mathbf{x}^1_{j+1}$,
print $\left(j + 1, \mathbf{x}_{j+1}, f\left(\mathbf{x}_{j+1}\right) \right)$,
goto step 11.

else

$h := \frac{h}{2}$,
goto step 3.

Step 11: (Termination condition)

If $j + 1 < \text{maxit}$,

then

$j := j + 1$,
goto step 1.

else

STOP.

The point

$$\mathbf{x}_s \in \{\mathbf{x}_0, \mathbf{x}_1, \ldots, \mathbf{x}_{\text{maxit}}\}$$

with the smallest function value is chosen as a starting point for a local minimization procedure.

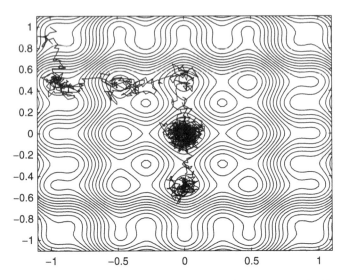

Fig. 3.5 Path of the randomized curve of steepest descent and contour lines, Problem 1, $\epsilon = 1$, 1,500 points

Now, we apply this algorithm to eight global minimization problems. The first problem serves as an example for the visualization of the numerical results.

Problem 1. $n = 2$ (Example 3.1)

$$\text{globmin}_{\mathbf{x}} \left\{ f : \mathbb{R}^2 \to \mathbb{R}, \, \mathbf{x} \mapsto 6x_1^2 + 6x_2^2 - \cos 12x_1 - \cos 12x_2 + 2 \right\},$$

This problem has 25 isolated minimum points within the box $[-1, 1] \times [-1, 1]$ with six different function values. Starting at $(-1, 1)^\top$ (very close to a local minimum point with the largest function value), Figs. 3.5–3.7 show a typical behavior of a path of the randomized curve of steepest descent (numerically approximated using the semi-implicit Euler method) in combination with

- Contour lines of the objective function
- Graph of the objective function
- Graph of the appropriate Lebesgue density function

Figure 3.8 shows a path with an ϵ too large.

This looks like purely random search using Gaussian distributed pseudorandom vectors without minimization part.

Now, we consider an ϵ too small (Figs. 3.9 and 3.10).

Figures 3.8, 3.9, and 3.10 show the importance of the possibility to match the free parameter ϵ to the global minimization problem.

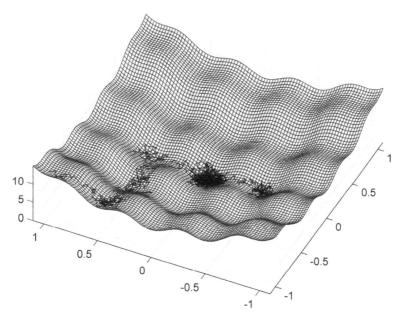

Fig. 3.6 Path of the randomized curve of steepest descent and objective function, Problem 1, $\epsilon = 1$, 1,500 points

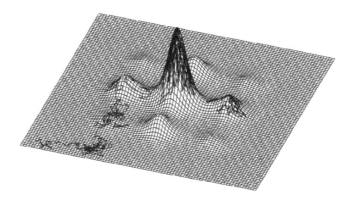

Fig. 3.7 Path of the randomized curve of steepest descent and Lebesgue density, Problem 1, $\epsilon = 1$, 1,500 points

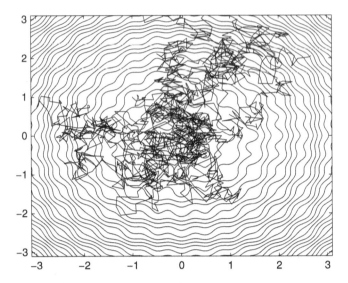

Fig. 3.8 Path of the randomized curve of steepest descent and contour lines, Problem 1, $\epsilon = 5$, 1,500 points

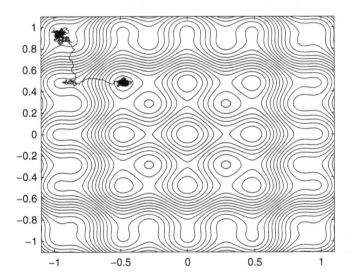

Fig. 3.9 Path of the randomized curve of steepest descent and contour lines, Problem 1, $\epsilon = 0.3$, 1,500 points

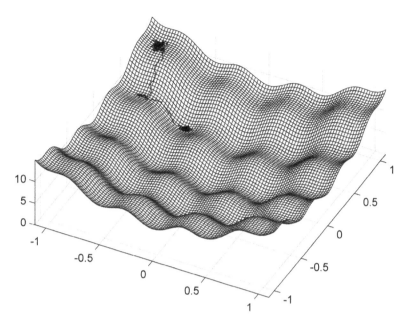

Fig. 3.10 Path of the randomized curve of steepest descent and objective function, Problem 1, $\epsilon = 0.3$, 1,500 points

Problem 2. $n = 90$

$$f : \mathbb{R}^{90} \to \mathbb{R}, \quad \mathbf{x} \mapsto 9.5x_1^2 - \cos(19x_1) + 792 \sum_{i=3}^{90} (3x_i + \sin^2(x_{i-1}))^2$$

$$+ \sqrt{3 + 12x_2^2 - 2\cos(12x_2) - 12x_2^2\cos(12x_2) + 36x_2^4 - \sin^2(12x_2)}$$

This objective function has 35 isolated minimum points with function values less than 17. The unique global minimum point is given by $\mathbf{x}_{gl} = \mathbf{0}$ with $f(\mathbf{x}_{gl}) = 0$.

$\epsilon = 5, \quad \delta = 0.1$.

Chosen number of iterations: maxit $= 500$.

$\mathbf{x}_0$ is chosen as a local minimum point with the largest function value.

$\mathbf{x}_s$ denotes the computed starting point for local minimization.

Results: $f(\mathbf{x}_s) = 0.1$.

Number of gradient evaluations: 1,246.

Number of Hessian evaluations: 1,246.

Local minimization leads to the global minimum point.

Problem 3. $n = 100$

$$f : \mathbb{R}^{100} \to \mathbb{R}, \quad \mathbf{x} \mapsto 6x_{100}^2 - \cos(12x_{100}) + 980 \sum_{i=1}^{97} (x_i - x_{i+2}^2)^2$$

$$+ \, 15x_{99}^2 - 2.5\cos(12x_{99}) + 18x_{98}^2 - 3\cos(12x_{98}) + 6.5$$

This objective function has 125 isolated minimum points with function values less than 21. The unique global minimum point is given by $\mathbf{x}_{gl} = \mathbf{0}$ with $f(\mathbf{x}_{gl}) = 0$.

$\epsilon = 3.5, \quad \delta = 0.1$.

Chosen number of iterations: maxit $= 3{,}000$.

$\mathbf{x}_0$ is chosen as a local minimum point with the largest function value.

$\mathbf{x}_s$ denotes the computed starting point for local minimization.

Results: $f(\mathbf{x}_s) = 0.27$.

Number of gradient evaluations: 6,160.

Number of Hessian evaluations: 6,160.

Local minimization leads to the global minimum point.

Now, we consider test problems from linear complementarity theory (see [Cottle.etal92]), which plays a very important role in game theory (computation of Nash equilibrium points (cf. [Owen68] and [Schäfer08])), and free boundary value problems (see [Crank84] and [Has.etal05]):

Given $\mathbf{c} \in \mathbb{R}^n$ and $\mathbf{C} \in \mathbb{R}^{n,n}$, find any $\mathbf{x} \in \mathbb{R}^n$ such that:

$$(\mathbf{c} + \mathbf{Cx})^\top \mathbf{x} = 0$$
$$x_i \geq 0 \quad i = 1, \ldots, n \qquad\qquad \text{(LCP)}$$
$$(\mathbf{c} + \mathbf{Cx})_i \geq 0 \quad i = 1, \ldots, n.$$

Using

$$P : \mathbb{R} \to \mathbb{R}, \quad x \mapsto \begin{cases} x & \text{for} \quad x > 0 \\ 0 & \text{for} \quad x \leq 0 \end{cases},$$

the first idea to solve (LCP) may consist in the investigation of unconstrained global optimization problems of the following type:

$$\operatorname*{globmin}_{\mathbf{x}} \left\{ \mathbf{c}^\top \mathbf{x} + \mathbf{x}^\top \mathbf{Cx} + \mu \left(\sum_{i=1}^n (P(-x_i))^4 + \sum_{i=1}^n (P(-(\mathbf{c} + \mathbf{Cx})_i))^4 \right) \right\},$$

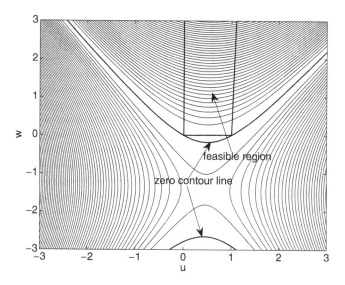

Fig. 3.11 Contour lines of f, LCP with two solutions, $n = 2$

where $\mu > 0$. Unfortunately, the objective function

$$g : \mathbb{R}^n \to \mathbb{R}, \quad \mathbf{x} \mapsto \mathbf{c}^\top \mathbf{x} + \mathbf{x}^\top \mathbf{C} \mathbf{x} + \mu \left(\sum_{i=1}^{n} (P(-x_i))^4 + \sum_{i=1}^{n} (P(-(\mathbf{c} + \mathbf{C}\mathbf{x})_i))^4 \right)$$

is not bounded from below in general. Therefore, we use the following objective function (again with $\mu > 0$):

$$f : \mathbb{R}^n \to \mathbb{R},$$

$$\mathbf{x} \mapsto \sqrt{1 + \left(\mathbf{c}^\top \mathbf{x} + \mathbf{x}^\top \mathbf{C} \mathbf{x} \right)^2} - 1 + \mu \left(\sum_{i=1}^{n} (P(-x_i))^4 + \sum_{i=1}^{n} (P(-(\mathbf{c} + \mathbf{C}\mathbf{x})_i))^4 \right),$$

with

- $f(\mathbf{x}) \geq 0$ for all $\mathbf{x} \in \mathbb{R}^n$.
- $\mathbf{x}^*$ solves (LCP) iff $f(\mathbf{x}^*) = 0$.

Example 3.7. We consider two examples with $n = 2$ visualized in Figs. 3.11–3.14:

The method of Best and Ritter [BesRit88] is chosen as a local minimization procedure of the quadratic problem

$$\operatorname*{locmin}_{\mathbf{x}} \{ \mathbf{c}^\top \mathbf{x} + \mathbf{x}^\top \mathbf{C} \mathbf{x}; \quad x_i \geq 0 \quad i = 1, \ldots, n$$

$$(\mathbf{c} + \mathbf{C}\mathbf{x})_i \geq 0 \quad i = 1, \ldots, n \}.$$

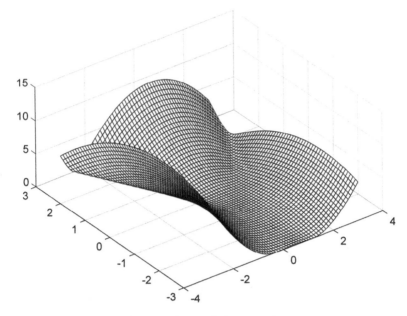

Fig. 3.12 Objective function f, LCP with two solutions, $n = 2$

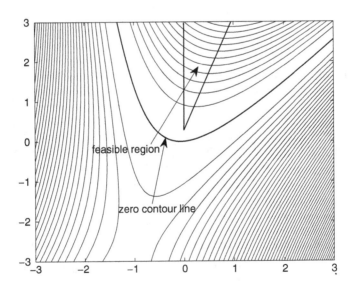

Fig. 3.13 Contour lines of f, LCP with no solution, $n = 2$

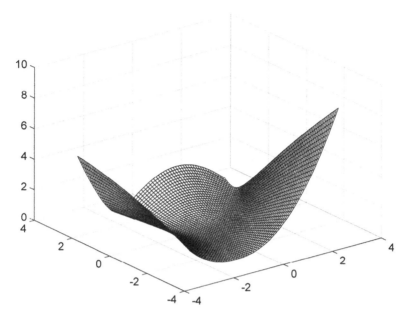

Fig. 3.14 Objective function f, LCP with no solution, $n = 2$

The following numerical results can be found in [Schä95] (with $\mu = 1{,}000$).

Problem 4. $n = 30$

$x^*_{2k} > 0$ is chosen randomly, $k = 1, \ldots, 15$.

$x^*_{2k-1} = 0, k = 1, \ldots, 15$.

$\mathbf{C} \in \mathbb{R}^{30,30}$ is computed by pseudorandom numbers.

$\mathbf{c}_{2k} = -(\mathbf{Cx}^*)_{2k}, k = 1, \ldots, 15$.

$\mathbf{c}_{2k-1}$ is chosen randomly such that $\mathbf{c}_{2k-1} \geq -(\mathbf{Cx}^*)_{2k-1}, k = 1, \ldots, 15$.

$\mathbf{x}_0$ is chosen far away from $\mathbf{x}^*$, $f(\mathbf{x}_0) = 2.8 \cdot 10^{10}$.

$\epsilon = 2, \quad \delta = 0.1$.

Chosen number of iterations: maxit $= 1{,}000$.

$\mathbf{x}_s$ denotes the computed starting point for local minimization.

$\mathbf{x}_{gl}$ denotes the computed global minimum point.

Results: $f(\mathbf{x}_s) = 20.7, f(\mathbf{x}_{gl}) = 6.7 \cdot 10^{-5}$.

Problem 5. $n = 40$

$x_{2k}^* > 0$ is chosen randomly, $k = 1, \ldots, 20$.

$x_{2k-1}^* = 0, k = 1, \ldots, 20$.

$C \in \mathbb{R}^{40,40}$ is computed by pseudorandom numbers.

$c_{2k} = -(Cx^*)_{2k}, k = 1, \ldots, 20$.

c_{2k-1} is chosen randomly such that $c_{2k-1} \geq -(Cx^*)_{2k-1}, k = 1, \ldots, 20$.

x_0 is chosen far away from x^*, $f(x_0) = 2.6 \cdot 10^{10}$.

$\epsilon = 2, \quad \delta = 0.1$.

Chosen number of iterations: maxit $= 1,000$.

x_s denotes the computed starting point for local minimization.

x_{gl} denotes the computed global minimum point.

Results: $f(x_s) = 31.6, f(x_{gl}) = 4.7 \cdot 10^{-5}$.

Problem 6. $n = 50$

$x_{2k}^* > 0$ is chosen randomly, $k = 1, \ldots, 25$.

$x_{2k-1}^* = 0, k = 1, \ldots, 25$.

$C \in \mathbb{R}^{50,50}$ is computed by pseudorandom numbers.

$c_{2k} = -(Cx^*)_{2k}, k = 1, \ldots, 25$.

c_{2k-1} is chosen randomly such that $c_{2k-1} \geq -(Cx^*)_{2k-1}, k = 1, \ldots, 25$.

x_0 is chosen far away from x^*, $f(x_0) = 2.3 \cdot 10^{10}$.

$\epsilon = 2, \quad \delta = 0.1$.

Chosen number of iterations: maxit $= 1,000$.

x_s denotes the computed starting point for local minimization.

x_{gl} denotes the computed global minimum point.

Results: $f(x_s) = 40.8, f(x_{gl}) = 8.9 \cdot 10^{-6}$.

Problem 7. $n = 60$

$x_{2k}^* > 0$ is chosen randomly, $k = 1, \ldots, 30$.

$x_{2k-1}^* = 0, k = 1, \ldots, 30$.

$C \in \mathbb{R}^{60,60}$ is computed by pseudorandom numbers.

$c_{2k} = -(\mathbf{Cx}^*)_{2k}, k = 1, \ldots, 30.$

c_{2k-1} is chosen randomly such that $c_{2k-1} \geq -(\mathbf{Cx}^*)_{2k-1}, k = 1, \ldots, 30.$

$\mathbf{x}_0$ is chosen far away from $\mathbf{x}^*$, $f(\mathbf{x}_0) = 3.8 \cdot 10^{10}$.

$\epsilon = 2, \quad \delta = 0.1.$

Chosen number of iterations: maxit = 1,000.

$\mathbf{x}_s$ denotes the computed starting point for local minimization.

$\mathbf{x}_{gl}$ denotes the computed global minimum point.

Results: $f(\mathbf{x}_s) = 54.2, f(\mathbf{x}_{gl}) = 2.7 \cdot 10^{-5}.$

Problem 8. $n = 70$

$x_{2k}^* > 0$ is chosen randomly, $k = 1, \ldots, 35.$

$x_{2k-1}^* = 0, k = 1, \ldots, 35.$

$\mathbf{C} \in \mathbb{R}^{70,70}$ is computed by pseudorandom numbers.

$c_{2k} = -(\mathbf{Cx}^*)_{2k}, k = 1, \ldots, 35.$

c_{2k-1} is chosen randomly such that $c_{2k-1} \geq -(\mathbf{Cx}^*)_{2k-1}, k = 1, \ldots, 35.$

$\mathbf{x}_0$ is chosen far away from $\mathbf{x}^*$, $f(\mathbf{x}_0) = 4.0 \cdot 10^{10}$.

$\epsilon = 2, \quad \delta = 0.1.$

Chosen number of iterations: maxit = 1,000.

$\mathbf{x}_s$ denotes the computed starting point for local minimization.

$\mathbf{x}_{gl}$ denotes the computed global minimum point.

Results: $f(\mathbf{x}_s) = 66.4, f(\mathbf{x}_{gl}) = 1.2 \cdot 10^{-5}.$

A collection of linear complementarity problems discussed in the literature is published in the *Handbook of Test Problems in Local and Global Optimization* ([Flo.etal99], Chap. 10). The maximal dimension of these problems is $n = 16$.

It is important to see that the efficiency of the semi-implicit Euler method apparently does not depend on the dimension of the problem. This is a very atypical property of a global optimization procedure.

On the other hand, the proposed approach may not work for unconstrained global minimization problems with a huge number of local minimum points as the following example shows.

Example 3.8. Considering

$$g_n : \mathbb{R}^n \to \mathbb{R}, \quad \mathbf{x} \mapsto \sum_{i=1}^{n} (4x_i^2 - \cos(8x_i) + 1),$$

Assumption 3.2 is fulfilled for all $\epsilon > 0$ and we know that each function g_n has 3^n isolated minimum points and a unique global minimum point at $\mathbf{x} = \mathbf{0}$ (cf. Example 1.1).

The region of attraction to the unique global minimum point is approximately given by $[-0.4, 0.4]^n$.

Choosing $\epsilon = \sqrt{2}$ and using the probability distribution given by the Lebesgue density function

$$\lambda_{g_n} : \mathbb{R}^n \to \mathbb{R}, \quad \mathbf{x} \mapsto \frac{\exp\left(-g_n(\mathbf{x})\right)}{\int\limits_{\mathbb{R}^n} \exp\left(-g_n(\mathbf{x})\right) d\mathbf{x}},$$

the probability for the region of attraction of the global minimum point is

$$p_{g_n} := \frac{\int\limits_{[-0.4, 0.4]^n} \exp\left(-g_n(\mathbf{x})\right) d\mathbf{x}}{\int\limits_{\mathbb{R}^n} \exp\left(-g_n(\mathbf{x})\right) d\mathbf{x}} \approx 0.8^n.$$

This probability decreases exponentially with the exponential growth of the number of isolated local minimum points.

3.4 An Euler Method with Gradient Approximations

In this section, we investigate an Euler method for the numerical solution of

$$\mathbf{X}_t(\tilde{\omega}) = \mathbf{x}_0 - \int_0^t \nabla f(\mathbf{X}_\tau(\tilde{\omega})) d\tau + \epsilon \left(\mathbf{B}_t(\tilde{\omega}) - \mathbf{B}_0(\tilde{\omega})\right), \quad t \in [0, \infty).$$

This Euler method is based on an approximation $\mathbf{x}_{\text{app}}(\bar{t}, \tilde{\omega})$ of $\mathbf{X}_{\bar{t}}(\tilde{\omega})$ and is given by

$$\mathbf{x}_{\text{app}}(\bar{t} + h, \tilde{\omega}) = \mathbf{x}_{\text{app}}(\bar{t}, \tilde{\omega}) - h\nabla f(\mathbf{x}_{\text{app}}(\bar{t}, \tilde{\omega})) + \epsilon \left(\mathbf{B}_{\bar{t}+h}(\tilde{\omega}) - \mathbf{B}_{\bar{t}}(\tilde{\omega})\right)$$

approximating the gradient by centered differences

$$\mathbf{D}f(\mathbf{x}(t, \tilde{\omega})) = \begin{pmatrix} \frac{f(\mathbf{x}(t,\tilde{\omega})_1 + \gamma, \mathbf{x}(t,\tilde{\omega})_2 \dots, \mathbf{x}(t,\tilde{\omega})_n) - f(\mathbf{x}(t,\tilde{\omega})_1 - \gamma, \mathbf{x}(t,\tilde{\omega})_2, \dots, \mathbf{x}(t,\tilde{\omega})_n)}{2\gamma} \\ \vdots \\ \frac{f(\mathbf{x}(t,\tilde{\omega})_1, \dots, \mathbf{x}(t,\tilde{\omega})_{n-1}, \mathbf{x}(t,\tilde{\omega})_n + \gamma) - f(\mathbf{x}(t,\tilde{\omega})_1, \dots, \mathbf{x}(t,\tilde{\omega})_{n-1}, \mathbf{x}(t,\tilde{\omega})_n - \gamma)}{2\gamma} \end{pmatrix}.$$

We arrive at the following algorithm, which uses only function values.

Step 0: (Initialization)

Choose $\mathbf{x}_0 \in \mathbb{R}^n$ and $\epsilon, \delta, \gamma > 0$,
Choose maxit $\in \mathbb{N}$,
$j := 0$,
goto step 1.

In step 0, the starting point $\mathbf{x}_0$, the parameter ϵ according to Assumption 3.2, the parameter $\delta > 0$ according to the step size control, the parameter $\gamma > 0$ according to the approximation of the gradient information, and the maximal number of iterations have to be determined by the user.

Step 1: (Gradient Approximation)

$h := 1$,
compute $\mathbf{D}f(\mathbf{x}_j)$
goto step 2.

The initial value $h_{\max}$ of the step size is chosen equal to 1.

Step 2: (Pseudorandom Numbers)

Compute $2n$ stochastically independent $\mathcal{N}(0, 1)$ Gaussian distributed pseudorandom numbers $p_1, \ldots p_n, q_1, \ldots, q_n \in \mathbb{R}$,
goto step 3.

In this step, the choice of the path is determined successively by the computer.

Step 3: (Computation of $\mathbf{x}_{j+1}^2$ by one step with step size h)

$$\mathbf{x}_{j+1}^2 := \mathbf{x}_j - h\mathbf{D}f(\mathbf{x}_j) + \epsilon\sqrt{\frac{h}{2}} \begin{pmatrix} p_1 + q_1 \\ \vdots \\ p_n + q_n \end{pmatrix}.$$

goto step 4.

$\mathbf{x}_{j+1}^2$ is computed by a step with starting point $\mathbf{x}_j$ using the step size h.

Step 4: (Computation of $\mathbf{x}_{\frac{h}{2}}$)

$$\mathbf{x}_{\frac{h}{2}} := \mathbf{x}_j - \frac{h}{2}\mathbf{D}f(\mathbf{x}_j) + \epsilon\sqrt{\frac{h}{2}} \begin{pmatrix} p_1 \\ \vdots \\ p_n \end{pmatrix}.$$

goto step 5.

$\mathbf{x}_{\frac{h}{2}}$ is computed by a step with starting point $\mathbf{x}_j$ using the step size $\frac{h}{2}$.

Step 5: (Gradient Approximation)

compute $\mathbf{D}f(\mathbf{x}_{\frac{h}{2}})$

goto step 6.

Step 6: (Computation of $\mathbf{x}^1_{j+1}$ by two steps with step size $\frac{h}{2}$)

$$\mathbf{x}^1_{j+1} := \mathbf{x}_{\frac{h}{2}} - \frac{h}{2}\mathbf{D}f(\mathbf{x}_{\frac{h}{2}}) + \epsilon\sqrt{\frac{h}{2}}\begin{pmatrix} q_1 \\ \vdots \\ q_n \end{pmatrix}.$$

goto step 7.

$\mathbf{x}^1_{j+1}$ is computed by a step with starting point $\mathbf{x}_{\frac{h}{2}}$ using the step size $\frac{h}{2}$.

Step 7: (Acceptance condition)

If $\left\| \mathbf{x}^1_{j+1} - \mathbf{x}^2_{j+1} \right\|_2 < \delta$,

then

        $\mathbf{x}_{j+1} := \mathbf{x}^1_{j+1}$,

        print $\left(j+1, \mathbf{x}_{j+1}, f\left(\mathbf{x}_{j+1} \right) \right)$,

        goto step 8.

else

        $h := \frac{h}{2}$,

        goto step 3.

Step 7: (Termination condition)

If $j+1 <$ maxit,

then

        $j := j+1$,

        goto step 1.

else

        STOP.

The point

$$\mathbf{x}_s \in \{\mathbf{x}_0, \mathbf{x}_1, \ldots, \mathbf{x}_{\text{maxit}}\}$$

with the smallest function value is chosen as a starting point for a local minimization procedure. The following numerical results are summarized in [Bar97].

Problem 9. $n = 80$

$$f : \mathbb{R}^{80} \to \mathbb{R}, \quad \mathbf{x} \mapsto 2 + 12x_{80}^2 - 2\cos(12x_{80}) + 720 \sum_{i=1}^{79} (x_i - \sin(\cos(x_{i+1}) - 1))^2.$$

This objective function has five isolated minimum points with function values less than 16. The unique global minimum point is given by $\mathbf{x}_{\mathrm{gl}} = \mathbf{0}$ with $f(\mathbf{x}_{\mathrm{gl}}) = 0$.

$\epsilon = 0.5, \quad \delta = 0.5, \quad \gamma = 10^{-6}$.

Chosen number of iterations: maxit $= 300$.

$\mathbf{x}_0 = (10, \ldots, 10)^\top$, $f(\mathbf{x}_0) = 1.1 \cdot 10^4$.

$\mathbf{x}_s$ denotes the computed starting point for local minimization.

Results: $f(\mathbf{x}_s) = 3.8565$, $\frac{\|\mathbf{x}_s\|_2}{80} = 0.024$.

Number of function evaluations: 144,781.

Local minimization leads to the global minimum point.

Problem 10. $n = 70$

$$f : \mathbb{R}^{70} \to \mathbb{R}, \quad \mathbf{x} \mapsto 1000 \sum_{i=2}^{70} (x_i - \ln(x_{i-1}^2 + 1))^2 - 1 +$$

$$+ \sqrt{3 + 19x_1^2 - 2\cos(19x_1) - 19x_1^2 \cos(19x_1) + 90.25x_1^4 - \sin^2(19x_1)}$$

This objective function has 7 isolated minimum points with function values less than 12. The unique global minimum point is given by $\mathbf{x}_{\mathrm{gl}} = \mathbf{0}$ with $f(\mathbf{x}_{\mathrm{gl}}) = 0$.

$\epsilon = 3.5, \quad \delta = 1.0, \quad \gamma = 10^{-6}$.

Chosen number of iterations: maxit $= 1,000$.

$\mathbf{x}_0 = (10, \ldots, 10)^\top$, $f(\mathbf{x}_0) = 2 \cdot 10^6$.

$\mathbf{x}_s$ denotes the computed starting point for local minimization.

Results: $f(\mathbf{x}_s) = 302.7$, $\frac{\|\mathbf{x}_s\|_2}{70} = 0.00811$.

Number of function evaluations: 422,681.

Local minimization leads to the global minimum point.

Chapter 4
Application: Optimal Decoding in Communications Engineering

4.1 Channel Coding

In digital communications, the transmission of information is technically realized by the transmission of vectors $\mathbf{u} \in \{\pm 1\}^k$ of information bits. Since such a transmission is physically disturbed by noise, one has to ensure that the probability of receiving a wrong value for u_i, $i = 1, \ldots, k$, is as small as possible. This can be achieved by the addition of $n - k$ redundant bits to $\mathbf{u} \in \{\pm 1\}^k$. Therefore, we transmit a vector $\mathbf{c} \in \mathcal{C} \subset \{\pm 1\}^n$ with $c_i = u_i, i = 1, \ldots, k$, where $\mathcal{C}$ denotes the set of all code words. For the choice of the additional $n - k$ components, the algebraic structure of $\{\pm 1\}$ given by the two commutative operations $\oplus$ and $\odot$ with

$$-1 \oplus -1 = +1$$
$$+1 \oplus +1 = +1$$
$$+1 \oplus -1 = -1$$
$$-1 \odot -1 = -1$$
$$+1 \odot -1 = +1$$
$$+1 \odot +1 = +1$$

is used. For each $i \in \{k + 1, \ldots, n\}$, a set $J_i \subseteq \{1, \ldots, k\}$ is chosen and c_i is computed by

$$c_i = \bigoplus_{j \in J_i} u_j, \qquad i = k + 1, \ldots, n.$$

The optimal choice of the number $(n - k)$ and of the sets $J_{k+1}, \ldots, J_n \subseteq \{1, \ldots, k\}$ is the purpose of channel coding (see, e.g., [vanLint98]). Let us consider a Hamming code with $k = 4$, $n = 7$, $J_5 = \{2, 3, 4\}$, $J_6 = \{1, 3, 4\}$, and $J_7 = \{1, 2, 4\}$, for instance. The information bit u_1 can be found triply in a code word $\mathbf{c} \in \mathcal{C}$:

  (i) Directly in $c_1 = u_1$
 (ii) Indirectly in $c_6 = u_1 \oplus u_3 \oplus u_4$
(iii) Indirectly in $c_7 = u_1 \oplus u_2 \oplus u_4$

S. Schäffler, *Global Optimization: A Stochastic Approach*, Springer Series in Operations Research and Financial Engineering, DOI 10.1007/978-1-4614-3927-1_4,
© Springer Science+Business Media New York 2012

The information bit u_2 can be found triply in a code word $\mathbf{c} \in \mathcal{C}$:

(i) Directly in $c_2 = u_2$
(ii) Indirectly in $c_5 = u_2 \oplus u_3 \oplus u_4$
(iii) Indirectly in $c_7 = u_1 \oplus u_2 \oplus u_4$

The information bit u_3 can be found triply in a code word $\mathbf{c} \in \mathcal{C}$:

(i) Directly in $c_3 = u_3$
(ii) Indirectly in $c_5 = u_2 \oplus u_3 \oplus u_4$
(iii) Indirectly in $c_6 = u_1 \oplus u_3 \oplus u_4$

Finally, the information bit u_4 can be found four times in a code word $\mathbf{c} \in \mathcal{C}$:

(i) Directly in $c_4 = u_4$,
(ii) Indirectly in $c_5 = u_2 \oplus u_3 \oplus u_4$
(iii) Indirectly in $c_6 = u_1 \oplus u_3 \oplus u_4$
(iv) Indirectly in $c_7 = u_1 \oplus u_2 \oplus u_4$

This Hamming code consists of the following $|\mathcal{C}| = 16$ code words:

$$
\begin{array}{ll}
+1 + 1 + 1 + 1 \mid +1 + 1 + 1 + 1 & +1 + 1 + 1 - 1 \mid -1 - 1 - 1 - 1 \\
+1 + 1 - 1 + 1 \mid -1 - 1 - 1 + 1 & +1 + 1 - 1 - 1 \mid +1 + 1 + 1 - 1 \\
+1 - 1 + 1 + 1 \mid -1 + 1 + 1 - 1 & +1 - 1 + 1 - 1 \mid +1 - 1 + 1 + 1 \\
+1 - 1 - 1 + 1 \mid +1 - 1 - 1 - 1 & +1 - 1 - 1 - 1 \mid -1 + 1 + 1 + 1 \\
-1 + 1 + 1 + 1 \mid +1 - 1 - 1 - 1 & -1 + 1 + 1 - 1 \mid -1 + 1 + 1 + 1 \\
-1 + 1 - 1 + 1 \mid -1 + 1 + 1 - 1 & -1 + 1 - 1 - 1 \mid +1 - 1 + 1 + 1 \\
-1 - 1 + 1 + 1 \mid -1 - 1 + 1 + 1 & -1 - 1 + 1 - 1 \mid +1 + 1 + 1 - 1 \\
-1 - 1 - 1 + 1 \mid +1 + 1 + 1 + 1 & -1 - 1 - 1 - 1 \mid -1 - 1 - 1 - 1
\end{array}
$$

A measure of the difference between two code words is the number of positions in which they differ. This measure is called

Hamming distance $d : \{\pm 1\}^n \times \{\pm 1\}^n \to \{0, \dots, n\}$.

The minimum distance $d_{\min}$ of a code is given by

$$
d_{\min} := \min_{\mathbf{c}_i, \mathbf{c}_j \in \mathcal{C}, \mathbf{c}_i \neq \mathbf{c}_j} \{d(\mathbf{c}_i, \mathbf{c}_j)\}.
$$

The minimum distance of our Hamming code is $d_{\min} = 3$. Therefore, if there occurs only one error during the transmission of a code word $\mathbf{c} \in \mathcal{C}$, this error can be corrected. In general, with the largest natural number L such that

$$L \le \frac{d_{\min} - 1}{2},$$

L errors in a code word can be corrected.

4.2 Decoding

The transmission of a binary vector $\mathbf{c} \in C$ leads to a received vector $\mathbf{y} \in \mathbb{R}^n$ caused by random noise. Depending on the stochastical properties of random noise, the problem consists in the reconstruction of the first k bits of $\mathbf{c}$ using the fact that the information contained in these first k bits is redundantly contained in the last $n - k$ bits of $\mathbf{c}$ also. There is a widely used channel model in digital communications, which is called AWGN model (**A**dditive **W**hite **G**aussian **N**oise), and which is given as follows (see [Proa95]).

The received vector $\mathbf{y} \in \mathbb{R}^n$ is a realization of an n-dimensional Gaussian distributed random variable $\mathbf{Y}$ with expectation

$$\mathbb{E}(\mathbf{Y}) = \mathbf{c}$$

and covariance matrix

$$\mathrm{Cov}(\mathbf{Y}) = \frac{n}{2k \cdot \mathrm{SNR}} \mathbf{I_n}$$

based on a probability space $(\Omega, \mathcal{S}, \mathbb{P})$. The positive constant SNR (**s**ignal to **n**oise **r**atio) represents the ratio between transmission energy for a single information bit and noise energy and is a measure for transmission costs. From source coding, we know that we can interpret $\mathbf{u} \in \{\pm 1\}^k$ as a realization of a random vector

$$\mathbf{U} : \Omega \to \{\pm 1\}^k$$

such that

- The component random variables $U_1, \ldots, U_k$ are stochastically independent.
- $\mathbb{P}(\{\omega \in \Omega; U_i(\omega) = +1\}) = \mathbb{P}(\{\omega \in \Omega; U_i(\omega) = -1\}) = \frac{1}{2}$ for all $i = 1, \ldots, k$.

It is not enough to reconstruct the values of $\mathbf{u} \in \{\pm 1\}^k$; furthermore, one is interested in a quantification of the quality of the decision, which can be done in the following way. Consider

$$L(i) := \ln \left(\frac{\mathbb{P}(\{\omega \in \Omega; U_i(\omega) = +1 | \mathbf{Y}(\omega) = y\})}{\mathbb{P}(\{\omega \in \Omega; U_i(\omega) = -1 | \mathbf{Y}(\omega) = y\})} \right), \quad i = 1, \ldots, k.$$

Of course,

$$L(i) > 0 \quad \text{leads to the decision } u_i = +1,$$
$$L(i) < 0 \quad \text{leads to the decision } u_i = -1,$$
$$L(i) = 0 \quad \text{no decision possible,}$$

where the quality of the descision is measured by $|L(i)|$. A mathematical analysis of the problem of the numerical computation of

$$\mathbf{x} := \begin{pmatrix} L(1) \\ \vdots \\ L(k) \end{pmatrix}$$

is done in [Schä97] and [Stu03]. It is shown that the computation of $\mathbf{x}$ leads to an unconstrained global minimization problem with objective function

$$f : \mathbb{R}^k \to \mathbb{R}, \quad \mathbf{x} \mapsto \sum_{i=1}^{k} \left(x_i - \frac{4k \cdot \text{SNR}}{n} y_i \right)^2$$

$$+ \sum_{i=k+1}^{n} \left(\ln \left(\frac{1 + \prod_{j \in J_i} \frac{\exp(x_j)-1}{\exp(x_j)+1}}{1 - \prod_{j \in J_i} \frac{\exp(x_j)-1}{\exp(x_j)+1}} \right) - \frac{4k \cdot \text{SNR}}{n} y_i \right)^2.$$

The objective function f is infinitely continuously differentiable, and Assumption 3.2 is fulfilled. For the following numerical analysis, we consider a widely used class of codes, namely, the BCH(n,k) (**B**ose–**C**haudhuri–**H**ocquenghem) codes (see [Proa95]). The information $\mathbf{u} \in \{\pm 1\}^k$ is chosen randomly. The transmission is simulated by Gaussian pseudorandom numbers (with predetermined SNR, measured in [dB]). We compare two decoding methods:

1. *BM method:* The first step consists in rounding the received vector
   $\mathbf{y} \in \mathbb{R}^n$:

$$\mathbf{y}_i > 0 \quad \Longrightarrow \quad \bar{c}_i = +1, \quad i = 1,\ldots,n$$
$$\mathbf{y}_i < 0 \quad \Longrightarrow \quad \bar{c}_i = -1, \quad i = 1,\ldots,n$$
$$\mathbf{y}_i = 0 \quad \Longrightarrow \quad \text{randomized decision.}$$

In the second step, a code word $\mathbf{c} \in \mathcal{C}$ of the used BCH(n,k) code with Hamming distance

$$d(\bar{\mathbf{c}}, \mathbf{c}) \le \frac{d_{\min} - 1}{2}$$

is searched. If this code word exists, it is unique, and $c_1, \ldots, c_k$ represents the result of decoding. If this code word does not exist, no decision is made. The BM method is still the standard technique used for this class of problems.

2. *Global optimization:* Application of the methods of Chap. 3 for the numerical solution of the unconstrained global minimization problem with objective function $f : \mathbb{R}^k \to \mathbb{R}$ yields a global minimum point $\mathbf{x}_{gl}$:

$$
\begin{aligned}
\mathbf{x}_{gl,i} > 0 &\implies u_i = +1, \quad i = 1, \ldots, k \\
\mathbf{x}_{gl,i} < 0 &\implies u_i = -1, \quad i = 1, \ldots, k \\
\mathbf{x}_{gl,i} = 0 &\implies \text{randomized decision.}
\end{aligned}
$$

The SNR was kept fixed until at least 100 decoding errors occurred. The ratio

$$
P_b := \frac{\text{number of decoding errors}}{\text{number of transmitted information bits}}
$$

versus SNR is shown in the following figures.

The global minimization approach is the best decoding method for BCH codes up to now. Figures 4.1–4.6 show that using global optimization, we obtain the same transmission quality (measured in P_b) at a lower SNR in comparison to the BM method. Note that a reduction of SNR by 1 dB means a reduction of 20% of transmission costs, a reduction of 2 dB of SNR means a reduction of 37% of transmission costs, and a reduction of 3 dB of SNR means a reduction of 50% of transmission costs. On the other hand, for fixed SNR, the global optimization method leads to a dramatic improvement of the transmission quality in comparison with the BM method. These effects are caused by the loss of information induced by rounding of the received vector $\mathbf{y} \in \mathbb{R}^n$ in the BM method.

The concatenation of codes is a widely used approach in source coding. In a first step, k_1 information bits $w_1, \ldots w_{k_1} \in \{\pm 1\}$ are encoded in a code word $\mathbf{u} \in \{\pm 1\}^k$, $k > k_1$ ($k - k_1$ redundant bits). In a second step, k bits $u_1, \ldots u_k \in \{\pm 1\}$ are encoded in a code word $\mathbf{c} \in \{\pm 1\}^n$, $n > k$ ($n - k$ redundant bits). Decoding the concatenation of codes via global optimization can be done in two steps:

1. *Decoding of the second code:* Using the received vector $\mathbf{y} \in \mathbb{R}^n$, find a global minimum point $\mathbf{x}_{gl}$ of the objective function

$$
f : \mathbb{R}^k \to \mathbb{R}, \quad \mathbf{x} \mapsto \sum_{i=1}^{k} \left(x_i - \frac{4k \cdot \text{SNR}}{n} y_i \right)^2
$$

$$
+ \sum_{i=k+1}^{n} \left(\ln \left(\frac{1 + \prod_{j \in J_i} \frac{\exp(x_j)-1}{\exp(x_j)+1}}{1 - \prod_{j \in J_i} \frac{\exp(x_j)-1}{\exp(x_j)+1}} \right) - \frac{4k \cdot \text{SNR}}{n} y_i \right)^2 .
$$

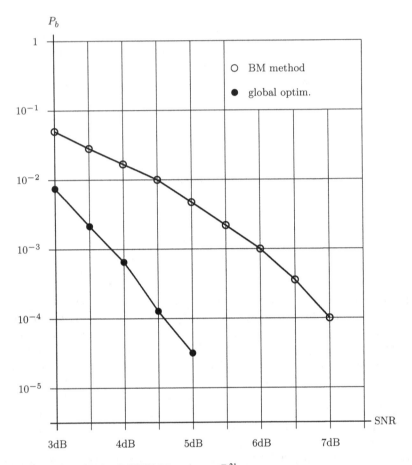

Fig. 4.1 Numerical results: BCH(31,21) code, $\mathbf{x} \in \mathbb{R}^{21}$

2. Interpreting the vector $\mathbf{x}_{gl}$ as a received vector after the first decoding step, the global minimum point $\mathbf{z}_{gl} \in \mathbb{R}^{k_1}$ of the objective function

$$f : \mathbb{R}^{k_1} \to \mathbb{R}, \quad \mathbf{z} \mapsto \sum_{i=1}^{k_1} \left(z_i - \frac{4k_1 \cdot \mathrm{SNR}_2}{k}(\mathbf{x}_{gl})_i \right)^2$$

$$+ \sum_{i=k_1+1}^{k} \left(\ln \left(\frac{1 + \prod\limits_{j \in J_i^*} \frac{\exp(z_j)-1}{\exp(z_j)+1}}{1 - \prod\limits_{j \in J_i^*} \frac{\exp(z_j)-1}{\exp(z_j)+1}} \right) - \frac{4k_1 \cdot \mathrm{SNR}_2}{k}(\mathbf{x}_{gl})_i \right)^2$$

is the result of decoding the first code and therefore the overall result of the decoding. The transmission of code word $\mathbf{c} \in \mathbb{R}^n$ with signal to noise ratio SNR

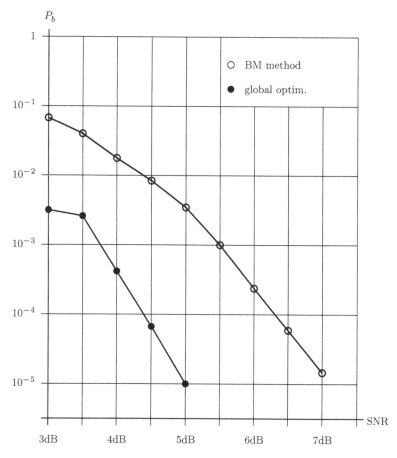

Fig. 4.2 Numerical results: BCH(63,45) code, $\mathbf{x} \in \mathbb{R}^{45}$

leads to a theoretical SNR_2 implicitly given by the empirical variance of $\mathbf{x}_{gl}$. The larger SNR_2 as compared to SNR, the better the first decoding step makes use of the information on bits $u_1, \ldots, u_k$ indirectly given in $c_{k+1}, \ldots, c_n$. Figures 4.7–4.12 show a comparison of SNR and SNR_2 under the assumption that

$$\frac{k_1}{k} = \frac{k}{n}$$

for the computed six examples.

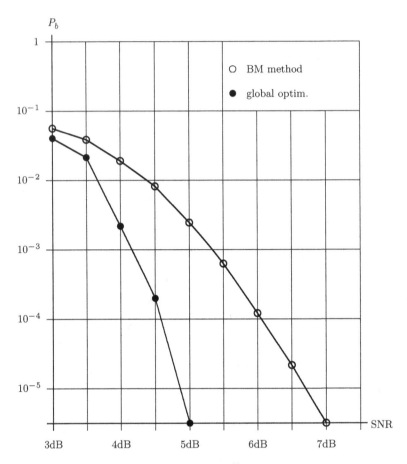

Fig. 4.3 Numerical results: BCH(127,99) code, $\mathbf{x} \in \mathbb{R}^{99}$

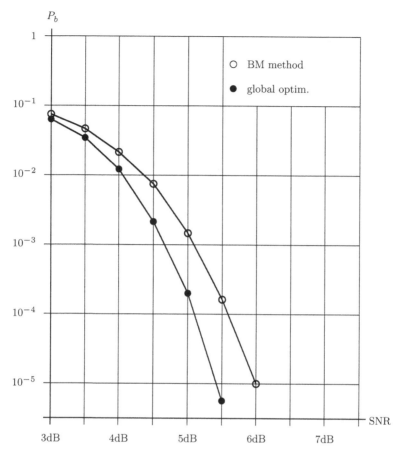

Fig. 4.4 Numerical results: BCH(255,191) code, $\mathbf{x} \in \mathbb{R}^{191}$

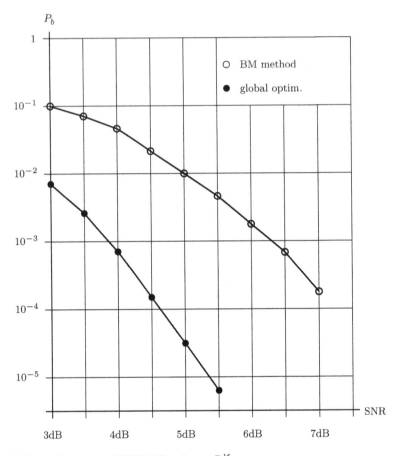

Fig. 4.5 Numerical results: BCH(31,16) code, $\mathbf{x} \in \mathbb{R}^{16}$

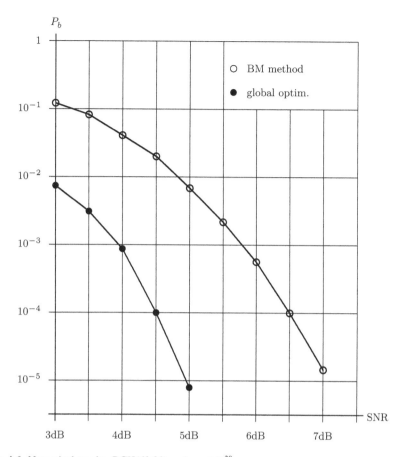

Fig. 4.6 Numerical results: BCH(63,30) code, $\mathbf{x} \in \mathbb{R}^{30}$

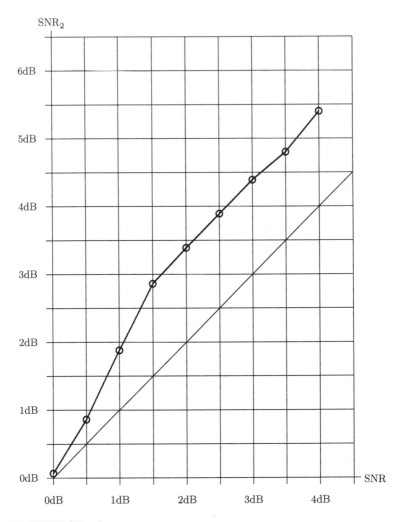

Fig. 4.7 BCH(31,21) code

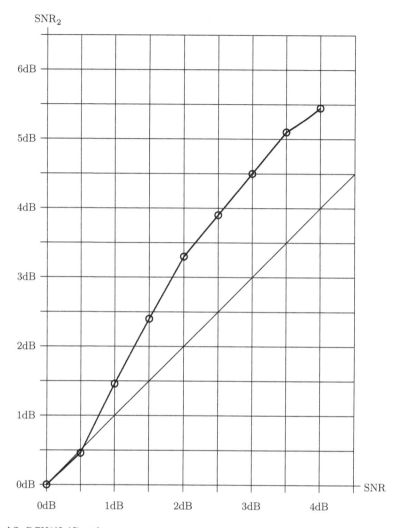

Fig. 4.8 BCH(63,45) code

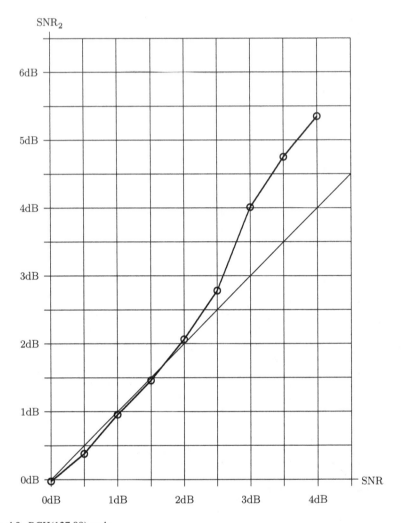

Fig. 4.9 BCH(127,99) code

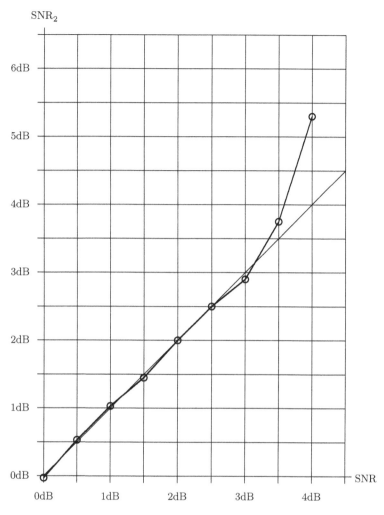

Fig. 4.10 BCH(255,191) code

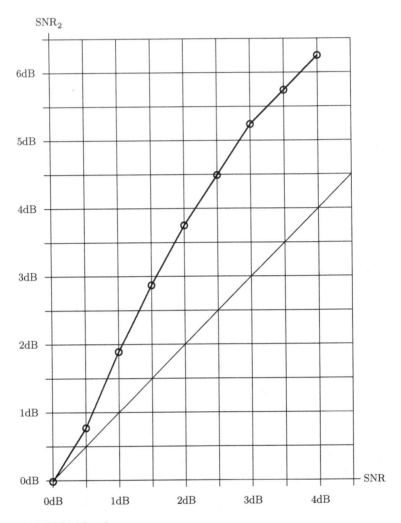

Fig. 4.11 BCH(31,16) code

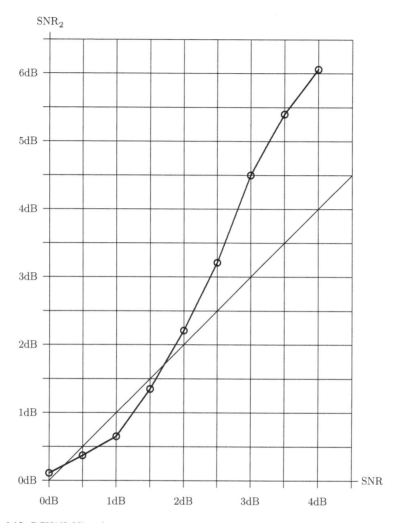

Fig. 4.12 BCH(63,30) code

Chapter 5
Constrained Global Optimization

5.1 Introduction

Now, we investigate constrained global minimization problems given as follows:

$$\begin{aligned}
\text{globmin}_{\mathbf{x}}\{ f(\mathbf{x}); \; & h_i(\mathbf{x}) = 0, \quad i = 1, \ldots, m \\
& h_i(\mathbf{x}) \leq 0, \quad i = m+1, \ldots, m+k \}, \\
& f, h_i : \mathbb{R}^n \to \mathbb{R}, \quad n \in \mathbb{N}, \; m, k \in \mathbb{N}_0, \\
& f, h_i \in C^2(\mathbb{R}^n, \mathbb{R}), \quad i = 1, \ldots, m+k.
\end{aligned}$$

We have to compute a point

$$\mathbf{x}_{gl} \in R := \{ \mathbf{x} \in \mathbb{R}^n; \; h_i(\mathbf{x}) = 0, \quad i = 1, \ldots, m$$
$$h_i(\mathbf{x}) \leq 0, \quad i = m+1, \ldots, m+k \}$$

such that

$$f(\mathbf{x}) \geq f(\mathbf{x}_{gl}) \quad \text{for all } \mathbf{x} \in R, \text{ where } R \neq \emptyset \text{ is assumed.}$$

Again, it is not reasonable to develop methods for solving constrained global minimization problems neglecting the existence of powerful methods solving constrained local minimization problems. Therefore, we are only interested in computing a suitable starting point close enough to a global minimum point in order to apply a local minimizing procedure. A constrained local optimization problem is given by

S. Schäffler, *Global Optimization: A Stochastic Approach*, Springer Series in Operations Research and Financial Engineering, DOI 10.1007/978-1-4614-3927-1_5, © Springer Science+Business Media New York 2012

$$\text{locmin}_{\mathbf{x}}\{f(\mathbf{x}); \ h_i(\mathbf{x}) = 0, \quad i = 1, \ldots, m$$

$$h_i(\mathbf{x}) \leq 0, \quad i = m+1, \ldots, m+k\},$$

$$f, h_i : \mathbb{R}^n \to \mathbb{R}, \quad n \in \mathbb{N}, \ m, k \in \mathbb{N}_0,$$

$$f, h_i \in C^2(\mathbb{R}^n, \mathbb{R}), \quad i = 1, \ldots, m+k,$$

i.e., one has to compute a point $\mathbf{x}_{\text{loc}} \in R$ such that

$$f(\mathbf{x}) \geq f(\mathbf{x}_{\text{loc}}) \quad \text{for all } \mathbf{x} \in R \cap U(\mathbf{x}_{\text{loc}}),$$

where $U(\mathbf{x}_{\text{loc}}) \subseteq \mathbb{R}^n$ is an open neighborhoood of $\mathbf{x}_{\text{loc}}$. Now, we consider the Karush–Kuhn–Tucker conditions for constrained minimization problems (see [Lue84]).

Karush–Kuhn–Tucker Conditions: Let $\mathbf{x}^* \in \mathbb{R}^n$ be a solution of

$$\text{locmin}_{\mathbf{x}}\{f(\mathbf{x}); \ h_i(\mathbf{x}) = 0, \quad i = 1, \ldots, m$$

$$h_i(\mathbf{x}) \leq 0, \quad i = m+1, \ldots, m+k\},$$

and let $J_{\mathbf{x}^*} = \{j_1, \ldots j_p\} \subseteq \{m+1, \ldots, m+k\}$ be the set of indices j for which

$$h_j(\mathbf{x}^*) = 0.$$

Assume that the gradient vectors of all active constraints in $\mathbf{x}^*$

$$\nabla h_1(\mathbf{x}^*), \ldots, \nabla h_m(\mathbf{x}^*), \nabla h_{j_1}(\mathbf{x}^*), \ldots, \nabla h_{j_p}(\mathbf{x}^*)$$

are linearly independent; then there exists a vector $\boldsymbol{\lambda} \in \mathbb{R}^m$ and a vector $\boldsymbol{\mu} \in \mathbb{R}^k$, $\mu_{m+1}, \ldots, \mu_{m+k} \geq 0$, such that

$$\nabla f(\mathbf{x}^*) + \sum_{i=1}^{m} \lambda_i \nabla h_i(\mathbf{x}^*) + \sum_{i=m+1}^{m+k} \mu_i \nabla h_i(\mathbf{x}^*) = 0$$

$$\sum_{i=m+1}^{m+k} \mu_i h_i(\mathbf{x}^*) = 0$$

$$h_i(\mathbf{x}^*) = 0, \quad i = 1, \ldots, m$$

$$h_i(\mathbf{x}^*) \leq 0, \quad i = m+1, \ldots, m+k.$$

Any $(\mathbf{x}, \boldsymbol{\lambda}, \boldsymbol{\mu}) \in \mathbb{R}^n \times \mathbb{R}^m \times \mathbb{R}^k_{\geq 0}$ fulfilling the Karush–Kuhn–Tucker conditions is called a Karush–Kuhn–Tucker point. Based on the Karush–Kuhn–Tucker conditions, a very popular method solving local constrained minimization problems is the SQP method.

The idea of the SQP method is to solve the above constrained local minimization problem by computing a sequence of Karush–Kuhn–Tucker points of special quadratic minimization problems with linear constraints. Assume

$$(\mathbf{x}^p, \boldsymbol{\lambda}^p, \boldsymbol{\mu}^p) \in \mathbb{R}^n \times \mathbb{R}^m \times \mathbb{R}^k_{\geq 0}, \quad p \in \mathbb{N},$$

then a point $(\mathbf{x}^{p+1}, \boldsymbol{\lambda}^{p+1}, \boldsymbol{\mu}^{p+1})$ is computed as a Karush–Kuhn–Tucker point of the local quadratic minimization problem:

$$
\operatorname*{locmin}_{\mathbf{x}} \Bigg\{ f(\mathbf{x}^p) + \nabla f(\mathbf{x}^p)^\top (\mathbf{x} - \mathbf{x}^p) + \frac{1}{2}(\mathbf{x} - \mathbf{x}^p)^\top \nabla^2 f(\mathbf{x}^p)(\mathbf{x} - \mathbf{x}^p)
$$

$$
+ \frac{1}{2}(\mathbf{x} - \mathbf{x}^p)^\top \left(\sum_{i=1}^{m} \lambda_i^p \nabla^2 h_i(\mathbf{x}^p) \right)(\mathbf{x} - \mathbf{x}^p)
$$

$$
+ \frac{1}{2}(\mathbf{x} - \mathbf{x}^p)^\top \left(\sum_{i=m+1}^{m+k} \mu_i^p \nabla^2 h_i(\mathbf{x}^p) \right)(\mathbf{x} - \mathbf{x}^p) ;
$$

$$
h_i(\mathbf{x}^p) + \nabla h_i(\mathbf{x}^p)^\top (\mathbf{x} - \mathbf{x}^p) = 0, \quad i = 1, \dots, m
$$

$$
h_i(\mathbf{x}^p) + \nabla h_i(\mathbf{x}^p)^\top (\mathbf{x} - \mathbf{x}^p) \leq 0, \quad i = m+1, \dots, m+k \Bigg\} .
$$

The starting point $(\mathbf{x}^0, \boldsymbol{\lambda}^0, \boldsymbol{\mu}^0) \in \mathbb{R}^n \times \mathbb{R}^m \times \mathbb{R}^k_{\geq 0}$ is chosen arbitrarily. A method for solving constraint global minimization problems using the SQP approach and pseudorandom numbers is proposed in [Schn11].

In this chapter, we are going to consider different approaches to solve constrained global optimization problems based on the two following methods for solving constrained local minimization problems:

1. *Penalty Methods*

The idea of penalty methods is to replace a constrained local minimization problem

$$
\operatorname*{locmin}_{\mathbf{x}} \{ f(\mathbf{x}); \ h_i(\mathbf{x}) = 0, \quad i = 1, \dots, m
$$

$$
h_i(\mathbf{x}) \leq 0, \quad i = m+1, \dots, m+k \},
$$

$$
f, h_i : \mathbb{R}^n \to \mathbb{R}, \quad n \in \mathbb{N}, \ m, k \in \mathbb{N}_0,
$$

$$
f, h_i \in C^2(\mathbb{R}^n, \mathbb{R}), \quad i = 1, \dots, m+k
$$

by a sequence of unconstrained global optimization problems with objective
function

$$f_{\text{penalty},c_p} : \mathbb{R}^n \to \mathbb{R}, \quad \mathbf{x} \mapsto f(\mathbf{x}) + c_p \left(\sum_{i=1}^{m} h_i(\mathbf{x})^4 + \sum_{i=m+1}^{m+k} (P(h_i(\mathbf{x})))^4 \right)$$

with

$$P : \mathbb{R} \to \mathbb{R}, \quad x \mapsto \begin{cases} x & \text{for} \quad x > 0 \\ 0 & \text{for} \quad x \le 0 \end{cases},$$

where $\{c_p\}_{p \in \mathbb{N}_0}$ is a sequence of positive, strictly monotonically increasing real
numbers with

$$\lim_{p \to \infty} c_p = \infty.$$

For large c_p, it is clear that a global minimum point $\mathbf{x}^*$ of the objective function
f_{penalty,c_p} (if it exists) will be in a region of points $\mathbf{x}$, for which

$$\left(\sum_{i=1}^{m} h_i(\mathbf{x})^4 + \sum_{i=m+1}^{m+k} \left(P(h_i(\mathbf{x}))^4 \right) \right)$$

is small. Unfortunately, with increasing penalty parameter c_p, the global minimization of f_{penalty,c_p} becomes more and more ill-conditioned. Therefore, it is not
advisable to start with a large penalty parameter c_0.

Example 5.1. Consider a local minimization problem with objective function

$$f : [6, 10] \times [0, 10] \to \mathbb{R}, \quad \mathbf{x} \mapsto 0.06x_1^2 + 0.06x_2^2 - \cos(1.2x_1) - \cos(1.2x_2) + 2.$$

The appropriate penalty functions are defined by

$$f_{\text{penalty},c_p} : \mathbb{R}^2 \to \mathbb{R}, \quad \mathbf{x} \mapsto 0.06x_1^2 + 0.06x_2^2 - \cos(1.2x_1) - \cos(1.2x_2) + 2$$

$$+ c_p \left((P(6 - x_1))^4 + (P(-x_2))^4 + (P(x_1 - 10))^4 + (P(x_2 - 10))^4 \right).$$

Figures 5.1–5.3 show the curve of steepest descent approximated by the semi-
implicit Euler method with

- Starting point $(2, 8)^\top$
- Penalty parameter $c_p = 0, 1, 100$
- Number of computed points: 1,500

In this case, the penalty approach leads to a local minimum point on the boundary
of the feasible region.

We will come back to this example in the next section.

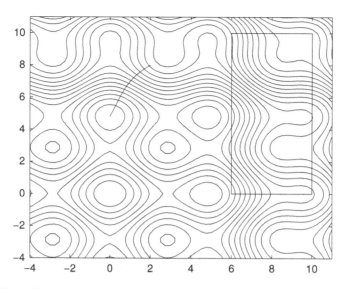

Fig. 5.1 Curve of steepest descent and contour lines, Example 5.1, $c_p = 0, 1, 500$ points

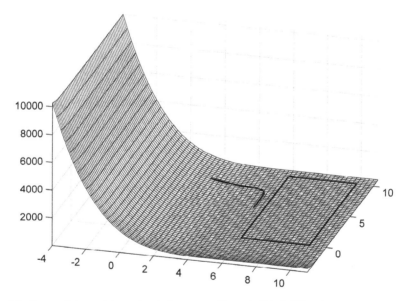

Fig. 5.2 Curve of steepest descent and $f_{\text{penalty},c_p=1}$, Example 5.1, 1, 500 points

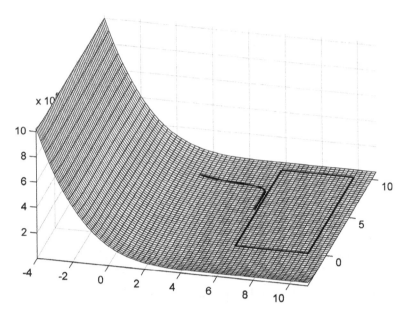

Fig. 5.3 Curve of steepest descent and $f_{\text{penalty},c_p=100}$, Example 5.1, 1, 500 points

2. *Equality Constraints*

Considering a local minimization problem with equality constraints

$$\operatorname*{locmin}_{\mathbf{x}}\{f(\mathbf{x});\ h_i(\mathbf{x}) = 0, \quad i = 1,\ldots,m\},$$

$$f, h_i : \mathbb{R}^n \to \mathbb{R}, \quad n \in \mathbb{N},\ m \in \mathbb{N}_0,$$

$$f, h_i \in C^2(\mathbb{R}^n, \mathbb{R}), \quad i = 1,\ldots,m,$$

it is possible to investigate the curve of steepest descent projected onto the differentiable $(n - m)$-dimensional manifold

$$M := \{\mathbf{x} \in \mathbb{R}^n;\ h_i(\mathbf{x}) = 0, \quad i = 1,\ldots,m\}$$

iff the gradient vectors

$$\nabla h_1(\mathbf{x}),\ldots,\nabla h_m(\mathbf{x})$$

are linearly independent for all $\mathbf{x} \in M$.

Using

$$\nabla \mathbf{h}(\mathbf{x}) := (\nabla h_1(\mathbf{x}),\ldots,\nabla h_m(\mathbf{x})) \in \mathbb{R}^{n,m}, \quad \mathbf{x} \in M,$$

the matrix

$$\mathbf{Pr}(\mathbf{x}) = \mathbf{I}_n - \nabla \mathbf{h}(\mathbf{x}) \left(\nabla \mathbf{h}(\mathbf{x})^\top \nabla \mathbf{h}(\mathbf{x})\right)^{-1} \nabla \mathbf{h}(\mathbf{x})^\top \in \mathbb{R}^{n,n}, \quad \mathbf{x} \in M,$$

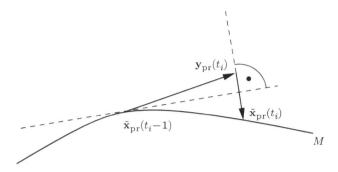

Fig. 5.4 Predictor–corrector method

represents the orthogonal projection onto the tangent space $T_{\mathbf{x}}M$ of M in $\mathbf{x} \in M$. Starting with a feasible starting point $\mathbf{x}_0 \in M$, which can be computed by solving an unconstrained global minimization problem with objective function

$$\bar{f} : \mathbb{R}^n \to \mathbb{R}, \quad \mathbf{x} \mapsto \sum_{i=1}^{m} h_i(\mathbf{x})^2$$

for instance, the projected curve of steepest descent

$$\mathbf{x}_{\mathrm{pr}} : [0, \infty) \to \mathbb{R}^n$$

is given by

$$\dot{\mathbf{x}}_{\mathrm{pr}}(t) = -\mathbf{Pr}(\mathbf{x}_{\mathrm{pr}}(t))\nabla f(\mathbf{x}_{\mathrm{pr}}(t)), \quad \mathbf{x}(0) = \mathbf{x}_0.$$

Since $\mathbf{x}_0 \in M$ and

$$\frac{d}{dt}h_i(\mathbf{x}_{\mathrm{pr}}(t)) = -\nabla h_i(\mathbf{x}_{\mathrm{pr}}(t))^\top \mathbf{Pr}(\mathbf{x}_{\mathrm{pr}}(t))\nabla f(\mathbf{x}_{\mathrm{pr}}(t))$$

$$= -\nabla f(\mathbf{x}_{\mathrm{pr}}(t))^\top \mathbf{Pr}(\mathbf{x}_{\mathrm{pr}}(t))\nabla h_i(\mathbf{x}_{\mathrm{pr}}(t))$$

$$= 0 \quad \text{for all } t \in [0, \infty), \quad i = 1, \dots, m,$$

we obtain

$$\mathbf{x}_{\mathrm{pr}}(t) \in M \quad \text{for all } t \in [0, \infty).$$

Properties of the projected curve of steepest descent analogous to Theorem 2.1 can be proven.

The numerical approximation of the projected curve of steepest descent consists of two steps (predictor–corrector method (Fig. 5.4)). Based on an approximation $\tilde{\mathbf{x}}_{\mathrm{pr}}(t_{i-1}) \in M$ of $\mathbf{x}_{\mathrm{pr}}(t_{i-1})$, a first approximation $\mathbf{y}_{\mathrm{pr}}(t_i)$ of $\mathbf{x}_{\mathrm{pr}}(t_i)$ is computed using numerical methods for ordinary differential equations (e.g., a semi-implicit Euler method). This is called predictor step. However, $\mathbf{y}_{\mathrm{pr}}(t_i) \notin M$ in general. Therefore, we need a second step to compute $\tilde{\mathbf{x}}_{\mathrm{pr}}(t_i) \in M$ based on $\mathbf{y}_{\mathrm{pr}}(t_i) \notin M$ (corrector step). Typically, the point $\tilde{\mathbf{x}}_{\mathrm{pr}}(t_i) \in M$ is assumed to be in the affine space

$$\left\{ \mathbf{x} \in \mathbb{R}^n;\ \mathbf{x} = \mathbf{y}_{\mathrm{pr}}(t_i) + \sum_{i=1}^{m} \alpha_i \nabla h_i(\mathbf{x}_{\mathrm{pr}}(t_{i-1})),\ \alpha_1, \dots, \alpha_m \in \mathbb{R} \right\},$$

which is orthogonal to the tangent space $T_{\tilde{\mathbf{x}}_{\mathrm{pr}}(t_{i-1})} M$ of M in $\tilde{\mathbf{x}}_{\mathrm{pr}}(t_{i-1}) \in M$.

Hence, the corrector step consists of the computation of a zero of the mapping

$$c : \mathbb{R}^m \to \mathbb{R}^m, \quad \boldsymbol{\alpha} \mapsto \begin{pmatrix} h_1 \left(\mathbf{y}_{\mathrm{pr}}(t_i) + \sum_{i=1}^{m} \alpha_i \nabla h_i(\mathbf{x}_{\mathrm{pr}}(t_{i-1})) \right) \\ \vdots \\ h_m \left(\mathbf{y}_{\mathrm{pr}}(t_i) + \sum_{i=1}^{m} \alpha_i \nabla h_i(\mathbf{x}_{\mathrm{pr}}(t_{i-1})) \right) \end{pmatrix}.$$

5.2   A Penalty Approach

Consider the constrained global minimization problem:

$$\operatorname*{globmin}_{\mathbf{x}}\{ f(\mathbf{x});\ h_i(\mathbf{x}) = 0, \quad i = 1, \dots, m$$

$$h_i(\mathbf{x}) \le 0, \quad i = m+1, \dots, m+k\},$$

$$f, h_i : \mathbb{R}^n \to \mathbb{R}, \quad n \in \mathbb{N},\ m, k \in \mathbb{N}_0,$$

$$f, h_i \in C^2(\mathbb{R}^n, \mathbb{R}), \quad i = 1, \dots, m+k$$

with feasible region

$$R = \{ \mathbf{x} \in \mathbb{R}^n;\ h_i(\mathbf{x}) = 0, \quad i = 1, \dots, m$$

$$h_i(\mathbf{x}) \le 0, \quad i = m+1, \dots, m+k\}.$$

Using

$$P : \mathbb{R} \to \mathbb{R}, \quad x \mapsto \begin{cases} x & \text{for } x > 0 \\ 0 & \text{for } x \le 0 \end{cases},$$

we investigate unconstrained global minimization problems with objective function

$$f_{\mathrm{penalty},c} : \mathbb{R}^n \to \mathbb{R}, \quad \mathbf{x} \mapsto f(\mathbf{x}) + c \left(\sum_{i=1}^{m} h_i(\mathbf{x})^4 + \sum_{i=m+1}^{m+k} (P(h_i(\mathbf{x})))^4 \right), \quad c > 0.$$

Now, we introduce the penalty approach proposed in [RitSch94]. In this paper, the main assumption is formulated as follows:

Assumption 5.2 *There exist real numbers $c_0, \epsilon, \rho > 0$ with:*

$$\mathbf{x}^\top \nabla f_{penalty,c}(\mathbf{x}) \geq \frac{1 + n\epsilon^2}{2} \max\{1, \|\nabla f_{penalty,c}(\mathbf{x})\|_2\}$$

for all $\mathbf{x} \in \{\mathbf{z} \in \mathbb{R}^n; \|\mathbf{z}\|_2 > \rho\}$ and for all $c \geq c_0$.

Assumption 5.2 means that Assumption 3.2 is fulfilled for all objective functions $f_{penalty,c}, c \geq c_0$, with the same $\epsilon, \rho > 0$.

In the following theorem, we study properties of the penalty approach.

Theorem 5.3 *Consider the constrained global minimization problem*

$$\text{globmin}_{\mathbf{x}}\{f(\mathbf{x}); \ h_i(\mathbf{x}) = 0, \quad i = 1, \ldots, m$$

$$h_i(\mathbf{x}) \leq 0, \quad i = m+1, \ldots, m+k\},$$

$$f, h_i : \mathbb{R}^n \to \mathbb{R}, \quad n \in \mathbb{N}, \ m, k \in \mathbb{N}_0,$$

$$f, h_i \in C^2(\mathbb{R}^n, \mathbb{R}), \quad i = 1, \ldots, m+k$$

and for each $c > 0$, the penalty functions

$$f_{penalty,c} : \mathbb{R}^n \to \mathbb{R}, \quad \mathbf{x} \mapsto f(\mathbf{x}) + c \left(\sum_{i=1}^{m} h_i(\mathbf{x})^4 + \sum_{i=m+1}^{m+k} (P(h_i(\mathbf{x})))^4 \right).$$

Let the following Assumption 5.2 be fulfilled:
There exist real numbers $c_0, \epsilon, \rho > 0$ with:

$$\mathbf{x}^\top \nabla f_{penalty,c}(\mathbf{x}) \geq \frac{1 + n\epsilon^2}{2} \max\{1, \|\nabla f_{penalty,c}(\mathbf{x})\|_2\}$$

for all $\mathbf{x} \in \{\mathbf{z} \in \mathbb{R}^n; \|\mathbf{z}\|_2 > \rho\}$ and for all $c \geq c_0$;
then we obtain:

(i) *The constrained global minimization problem has a solution.*
(ii) *For any sequence $\{c_p\}_{p \in \mathbb{N}_0}$ with*

- $c_{p+1} > c_p > c_0$ *for all $p \in \mathbb{N}$,*
- $\lim_{p \to \infty} c_p = \infty$,

let $\mathbf{x}_p^*$ be a global minimum point of $f_{penalty,c_p}$ (the existence is guaranteed by Assumption 5.2). Then the sequence $\{\mathbf{x}_p^*\}_{p \in \mathbb{N}_0}$ has at least one cluster point, and each cluster point is a global minimum point of the constrained global minimization problem.

Proof. Consider $\{\mathbf{x}_p^*\}_{p\in\mathbb{N}_0}$. Assumption 5.2 guarantees that

$$\mathbf{x}_p^* \in \{\mathbf{z} \in \mathbb{R}^n;\ \|\mathbf{z}\|_2 \le \rho\}.$$

Therefore, each sequence $\{\mathbf{x}_p^*\}_{p\in\mathbb{N}_0}$ has at least one cluster point denoted by $\mathbf{x}_{cl}^*$. Let $\{\mathbf{x}_{p_l}^*\}_{l\in\mathbb{N}}$ be a subsequence of $\{\mathbf{x}_p^*\}_{p\in\mathbb{N}_0}$ with

$$p_i > p_j \quad \text{for all } i > j$$

and with

$$\lim_{l\to\infty} \mathbf{x}_{p_l}^* = \mathbf{x}_{cl}^*.$$

Assume that

$$\mathbf{x}_{cl}^* \notin R,$$

then we get from the definition of $\{c_p\}_{p\in\mathbb{N}_0}$ and from the fact that R is a closed subset of $\mathbb{R}^n$:

$$\lim_{l\to\infty} f_{\text{penalty},c_{p_l}}(\mathbf{x}_{p_l}^*) = \infty.$$

On the other hand, we know that

$$f_{\text{penalty},c_{p_l}}(\mathbf{x}_{p_l}^*) \le \inf_{\mathbf{x}\in R}\{f(\mathbf{x})\} \quad \text{for all } l \in \mathbb{N}.$$

This contradiction leads to

$$\mathbf{x}_{cl}^* \in R.$$

Assume that

$$f(\mathbf{x}_{cl}^*) > \inf_{\mathbf{x}\in R}\{f(\mathbf{x})\},$$

then we obtain

$$f_{\text{penalty},c_{p_l}}(\mathbf{x}_{cl}^*) > \inf_{\mathbf{x}\in R}\{f(\mathbf{x})\} \quad \text{for all } l \in \mathbb{N}.$$

Otherwise, we get from $f_{\text{penalty},c_{p_l}}(\mathbf{x}_{p_l}^*) \le \inf_{\mathbf{x}\in R}\{f(\mathbf{x})\}$ for all $l \in \mathbb{N}$:

$$f_{\text{penalty},c_{p_l}}(\mathbf{x}_{cl}^*) = f_{\text{penalty},c_{p_l}}\left(\lim_{i\to\infty} \mathbf{x}_{p_i}^*\right) \le \inf_{\mathbf{x}\in R}\{f(\mathbf{x})\} \quad \text{for all } l \in \mathbb{N}.$$

This contradiction leads to

$$f(\mathbf{x}_{cl}^*) = \inf_{\mathbf{x}\in R}\{f(\mathbf{x})\}.$$

q.e.d.

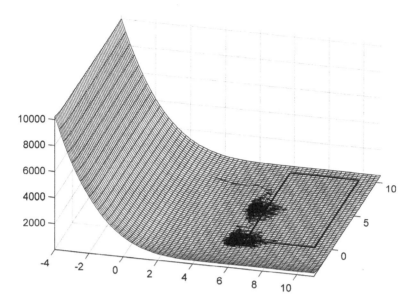

Fig. 5.5 A path of the randomized curve of steepest descent and $f_{\text{penalty},c_p=1}$, $\epsilon = 1$, $1,500$ points

Now, we come back to Example 5.1 and consider

Problem 11. $n = 2$

$$f : [6, 10] \times [0, 10] \rightarrow \mathbb{R}, \quad \mathbf{x} \mapsto 0.06x_1^2 + 0.06x_2^2 - \cos(1.2x_1) - \cos(1.2x_2) + 2$$

with the appropriate penalty functions

$$f_{\text{penalty},c_p} : \mathbb{R}^2 \rightarrow \mathbb{R}, \quad \mathbf{x} \mapsto 0.06x_1^2 + 0.06x_2^2 - \cos(1.2x_1) - \cos(1.2x_2) + 2$$
$$+ c_p \left((P(6 - x_1))^4 + (P(-x_2))^4 + (P(x_1 - 10))^4 + (P(x_2 - 10))^4 \right).$$

The function f has one unique global minimum point at $(6, 0)^\top$. Assumption 5.2 is fulfilled for all $\epsilon > 0$. Figures 5.5 and 5.6 show 1,500 points of paths of the randomized curve of steepest descent with $\epsilon = 1$, starting point $(2, 8)^\top$, and $c_p = 1,100$, respectively.

In Sect. 3.3, we have considered linear complementarity problems (LCP) of the following type:

Given $\mathbf{c} \in \mathbb{R}^n$ and $\mathbf{C} \in \mathbb{R}^{n,n}$, find any $\mathbf{x} \in \mathbb{R}^n$ such that

$$(\mathbf{c} + \mathbf{Cx})^\top \mathbf{x} = 0$$
$$x_i \geq 0 \quad i = 1, \ldots, n$$
$$(\mathbf{c} + \mathbf{Cx})_i \geq 0 \quad i = 1, \ldots, n.$$

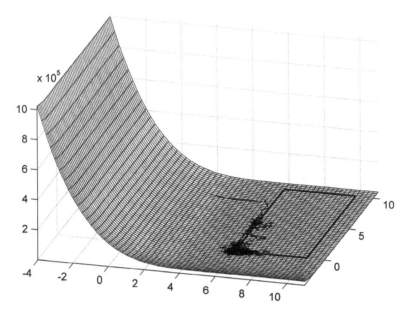

Fig. 5.6 A path of the randomized curve of steepest descent and $f_{\text{penalty},c_p=100}$, $\epsilon = 1$, $1,500$ points

Each solution of a given (LCP) problem is a solution of the appropriate constrained global minimization problem

$$\operatorname*{globmin}_{\mathbf{x}} \left\{ \sqrt{1 + \left(\mathbf{c}^{\top}\mathbf{x} + \mathbf{x}^{\top}\mathbf{C}\mathbf{x}\right)^2} - 1; \quad -x_i \leq 0, \quad i = 1, \ldots, n, \right.$$

$$\left. (-\mathbf{c} - \mathbf{C}\mathbf{x})_i \leq 0, \quad i = 1, \ldots, n \right\}$$

with objective function value equal to zero and vice versa. Using a penalty approach with $c = 1,000$, one has to solve an unconstrained global minimization problem with objective function

$$f : \mathbb{R}^n \to \mathbb{R}, \quad \mathbf{x} \mapsto \sqrt{1 + \left(\mathbf{c}^{\top}\mathbf{x} + \mathbf{x}^{\top}\mathbf{C}\mathbf{x}\right)^2} - 1$$

$$+ 1000 \left(\sum_{i=1}^{n} \left(P(-x_i)\right)^4 + \sum_{i=1}^{n} \left(P(-(\mathbf{c} + \mathbf{C}\mathbf{x})_i)\right)^4 \right),$$

which was done in Sect. 3.3, Problems 4–8.

5.3 Equality Constraints

In this section, we investigate global minimization problems with equality constraints:

$$\text{globmin}_{\mathbf{x}}\{f(\mathbf{x}); \ h_i(\mathbf{x}) = 0, \quad i = 1, \dots, m\},$$

$$f, h_i : \mathbb{R}^n \to \mathbb{R}, \quad n \in \mathbb{N}, \ m \in \mathbb{N}_0,$$

$$f, h_i \in C^2(\mathbb{R}^n, \mathbb{R}), \quad i = 1, \dots, m.$$

Again, we assume that

$$M := \{\mathbf{x} \in \mathbb{R}^n; \ h_i(\mathbf{x}) = 0, \quad i = 1, \dots, m\}$$

is a differentiable $(n - m)$-dimensional manifold, which is equivalent to the postulation that the gradient vectors

$$\nabla h_1(\mathbf{x}), \dots, \nabla h_m(\mathbf{x})$$

are linearly independent for all $\mathbf{x} \in M$.

Using

$$\nabla \mathbf{h}(\mathbf{x}) := (\nabla h_1(\mathbf{x}), \dots, \nabla h_m(\mathbf{x})) \in \mathbb{R}^{n,m}, \quad \mathbf{x} \in M,$$

using the projection matrix

$$\mathbf{Pr}(\mathbf{x}) = \mathbf{I}_n - \nabla \mathbf{h}(\mathbf{x}) \left(\nabla \mathbf{h}(\mathbf{x})^\top \nabla \mathbf{h}(\mathbf{x})\right)^{-1} \nabla \mathbf{h}(\mathbf{x})^\top \in \mathbb{R}^{n,n}$$

onto the tangent space $T_\mathbf{x} M$ of M in $\mathbf{x} \in M$, and using a feasible starting point $\mathbf{x}_0 \in M$, we have considered the projected curve of steepest descent

$$\dot{\mathbf{x}}_{\text{pr}}(t) = -\mathbf{Pr}(\mathbf{x}_{\text{pr}}(t))\nabla f(\mathbf{x}_{\text{pr}}(t)), \quad \mathbf{x}(0) = \mathbf{x}_0,$$

for local minimization. Analogously to the unconstrained case, we try to solve global minimization problems with equality constraints by a suitable randomization of the projected curve of steepest descent. Based on the Wiener space $(\Omega, \mathcal{B}(\Omega), W)$ and an n-dimensional Brownian Motion $\{\mathbf{B}_t\}_{t \in [0,\infty)}$, the first idea may consist in solving

$$\bar{\mathbf{X}}_{\text{pr}}(t, \omega) = \mathbf{x}_0 - \int_0^t \mathbf{Pr}(\bar{\mathbf{X}}_{\text{pr}}(\tau, \omega))\nabla f(\bar{\mathbf{X}}_{\text{pr}}(\tau, \omega))d\tau + \epsilon\,(\mathbf{B}_t(\omega) - \mathbf{B}_0(\omega))$$

with $t \in [0, \infty)$ and $\omega \in \Omega$, where the role of ϵ has to be clarified. Unfortunately, $\bar{\mathbf{X}}_{\text{pr}}(t, \omega) \notin M$ in general for each $\epsilon > 0$, $t \in (0, \infty)$, and $\omega \in \Omega$. Therefore, we have to look for a stochastic process $\{\mathbf{S}_t\}_{t \in [0,\infty)}$ for some $\epsilon > 0$ based on $(\Omega, \mathcal{B}(\Omega), W)$ such that we obtain results analogously to the unconstrained case using

$$\mathbf{X}_{\mathrm{pr}}(t, \omega) = \mathbf{x}_0 - \int\limits_0^t \mathbf{Pr}(\mathbf{X}_{\mathrm{pr}}(\tau, \omega))\nabla f(\mathbf{X}_{\mathrm{pr}}(\tau, \omega))d\tau + \epsilon \mathbf{S}_t(\omega)$$

and such that we obtain in particular $\mathbf{X}_{\mathrm{pr}}(t, \omega) \in M$, $t \in [0, \infty)$, and $\omega \in \Omega$. This problem leads to the Fisk–Stratonovich integral in stochastic analysis, which we introduce now without going into technical details (see [Pro95], for instance).

Let

$$\mathbf{Y}_t : \Omega \to \mathbb{R}^{n,n}, \quad t \in [0, \infty),$$

be a matrix-valued stochastic process defined on $(\Omega, \mathcal{B}(\Omega), W)$ and consider a sequence $\left\{t_0^{(q)}, t_1^{(q)}, \ldots, t_{p_q}^{(q)}\right\}_{q \in \mathbb{N}}$ of discretizations of the interval $[0, T]$, $T > 0$ such that

- $p_i \in \mathbb{N}$ for all $i \in \mathbb{N}$,
- $p_i < p_j$ for $i < j$,
- $0 = t_0^{(q)} < t_1^{(q)} < \ldots < t_{p_q}^{(q)} = T$ for all $q \in \mathbb{N}$,
- $\lim\limits_{q \to \infty} \left(\max\limits_{i=1,\ldots,p_q} \left(t_i^{(q)} - t_{i-1}^{(q)} \right) \right) = 0$,

then we define the Fisk–Stratonovich integral of $\{\mathbf{Y}_t\}_{t \in [0,\infty)}$ with respect to the Brownian Motion $\{\mathbf{B}_t\}_{t \in [0,\infty)}$ by

$$\int\limits_0^T \mathbf{Y}_t \circ d\mathbf{B}_t := L^2\text{-}\lim\limits_{q \to \infty} \sum_{i=1}^{p_q} \mathbf{Y}_{\frac{t_i^{(q)} + t_{i-1}^{(q)}}{2}} \left(\mathbf{B}_{t_i^{(q)}} - \mathbf{B}_{t_{i-1}^{(q)}} \right),$$

where L^2-lim denotes the component-by-component L^2-convergence defined as follows:

Let $\{\xi_i\}_{i \in \mathbb{N}}$ be a real-valued stochastic process defined on the Wiener space $(\Omega, \mathcal{B}(\Omega), W)$ such that

$$\int \xi_i^2 dW < \infty, \quad i \in \mathbb{N},$$

and let ξ be a real-valued random variable defined on the Wiener space $(\Omega, \mathcal{B}(\Omega), W)$ such that

$$\int \xi^2 dW < \infty,$$

then

$$L^2\text{-}\lim\limits_{i \to \infty} \xi_i = \xi \quad :\Longleftrightarrow \quad \lim\limits_{i \to \infty} \int (\xi - \xi_i)^2 dW = 0.$$

Note that $\int\limits_0^T \mathbf{Y}_t \circ d\mathbf{B}_t$ is an n-dimensional random variable on $(\Omega, \mathcal{B}(\Omega), W)$. It is very important to recognize that in the formula

$$L^2\text{-}\lim_{q\to\infty} \sum_{i=1}^{p_q} \mathbf{Y}_{\frac{t_i^{(q)}+t_{i-1}^{(q)}}{2}} \left(\mathbf{B}_{t_i^{(q)}} - \mathbf{B}_{t_{i-1}^{(q)}}\right)$$

one has to evaluate $\{\mathbf{Y}_t\}_{t\in[0,\infty)}$ exactly at the midpoint $\frac{t_i^{(q)}+t_{i-1}^{(q)}}{2}$ of the interval $\left[t_{i-1}^{(q)}, t_i^{(q)}\right]$. Evaluation at $t_{i-1}^{(q)}$ instead of $\frac{t_i^{(q)}+t_{i-1}^{(q)}}{2}$ leads to the well-known Itô integral, which is widely used in mathematical finance. But, as shown in [Elw82], only the Fisk–Stratonovich integral guarantees that

$$\mathbf{X}_{\mathrm{pr}}(\tau, \omega) \in M, \quad \tau \in [0, t], \quad \omega \in \Omega,$$

where

$$\mathbf{X}_{\mathrm{pr}}(t, \omega) = \mathbf{x}_0 - \int_0^t \mathbf{Pr}(\mathbf{X}_{\mathrm{pr}}(\tau, \omega)) \nabla f(\mathbf{X}_{\mathrm{pr}}(\tau, \omega)) d\tau$$

$$+ \epsilon \underbrace{\int_0^t \mathbf{Pr}(\mathbf{X}_{\mathrm{pr}}(\tau, \bullet)) \circ d\mathbf{B}_\tau(\omega)}_{=:\,\mathbf{S}_t(\omega)}, \quad \omega \in \Omega,$$

and where

$$\mathbf{X}_{\mathrm{pr}}(\tau, \bullet) : \Omega \to \mathbb{R}^n, \quad \omega \mapsto \mathbf{X}_{\mathrm{pr}}(\tau, \omega).$$

Hence, we have found a suitable randomization of the projected curve of steepest descent analogously to the unconstrained case. It is shown in [Stö00] that this approach preserves the properties of the unconstrained case as we summarize in the following.

It is possible to define a canonical metric

$$d_M : M \times M \to \mathbb{R}$$

on M, which induces the canonical inner product in $\mathbb{R}^{n-m}$ on the tangent spaces $T_{\mathbf{x}}M$ of M in $\mathbf{x} \in M$. Using this metric, we have geodetic lines on M, which play the role of straight lines in $\mathbb{R}^{n-m}$, and we have balls

$$B(\mathbf{x}_0, r) := \{\mathbf{x} \in M;\, d_M(\mathbf{x}, \mathbf{x}_0) \le \rho\}, \quad \rho > 0, \quad \mathbf{x}_0 \in M$$

in M. Therefore, we can try to translate Assumption 3.2 informally to an equivalent assumption for the manifold M:

Fix $\mathbf{x}_0 \in M$. The objective function f has to increase sufficiently fast (growth controlled by some $\epsilon > 0$) along each geodetic line starting at $\mathbf{x}_0$ outside a ball $\{\mathbf{x} \in M; d_M(\mathbf{x}, \mathbf{x}_0) \leq \rho\}$, where only the existence of $\rho > 0$ is postulated.

A mathematically accurate formulation of an analogon of this informal statement needs a lot of preliminaries from differential geometry and from the theory of stochastic differential equations of Fisk–Stratonovich type and is given in [Stö00], where results analogous to those of Theorems 3.3 and 3.6 are shown for the equality constrained case and where the following numerical analysis is derived.

Since

$$
\int_0^T \mathbf{Y}_t \circ d\mathbf{B}_t = L^2\text{-}\lim_{q \to \infty} \sum_{i=1}^{p_q} \mathbf{Y}_{\frac{t_i^{(q)} + t_{i-1}^{(q)}}{2}} \left(\mathbf{B}_{t_i^{(q)}} - \mathbf{B}_{t_{i-1}^{(q)}} \right)
$$

$$
= L^2\text{-}\lim_{q \to \infty} \sum_{i=1}^{p_q} \frac{1}{2} \left(\mathbf{Y}_{t_i^{(q)}} + \mathbf{Y}_{t_{i-1}^{(q)}} \right) \left(\mathbf{B}_{t_i^{(q)}} - \mathbf{B}_{t_{i-1}^{(q)}} \right)
$$

for all stochastic processes $\mathbf{Y}$ with continuous paths, it is obvious to approximate the two integrals in

$$
\mathbf{X}_{\text{pr}}(t, \omega) = \mathbf{x}_0 - \int_0^t \mathbf{Pr}(\mathbf{X}_{\text{pr}}(\tau, \omega)) \nabla f(\mathbf{X}_{\text{pr}}(\tau, \omega)) d\tau
$$

$$
+ \epsilon \int_0^t \mathbf{Pr}(\mathbf{X}_{\text{pr}}(\tau, \bullet)) \circ d\mathbf{B}_\tau(\omega)
$$

by the trapezoidal rule:

$$
\int_{\bar{t}}^{\bar{t}+h} \mathbf{Pr}(\mathbf{X}_{\text{pr}}(\tau, \omega)) \nabla f(\mathbf{X}_{\text{pr}}(\tau, \omega)) d\tau \approx \frac{h}{2} \Bigg(\mathbf{Pr}(\mathbf{X}_{\text{pr}}(\bar{t}, \omega)) \nabla f(\mathbf{X}_{\text{pr}}(\bar{t}, \omega))
$$

$$
+ \mathbf{Pr}(\mathbf{X}_{\text{pr}}(\bar{t} + h, \omega)) \nabla f(\mathbf{X}_{\text{pr}}(\bar{t} + h, \omega)) \Bigg),
$$

$$
\int_{\bar{t}}^{\bar{t}+h} \mathbf{Pr}(\mathbf{X}_{\text{pr}}(\tau, \bullet)) \circ d\mathbf{B}_\tau(\omega) \approx \frac{1}{2} \Bigg(\mathbf{Pr}(\mathbf{X}_{\text{pr}}(\bar{t}, \omega)) + \mathbf{Pr}(\mathbf{X}_{\text{pr}}(\bar{t} + h, \omega)) \Bigg) \cdot
$$

$$
\cdot \left(\mathbf{B}_{\bar{t}+h}(\omega) - \mathbf{B}_{\bar{t}}(\omega) \right).
$$

Based on an approximation $\mathbf{x}_{\text{app}}(\bar{t}, \tilde{\omega})$ of $\mathbf{X}_{\text{pr}}(\bar{t}, \tilde{\omega})$, the trapezoidal rule for the predictor step leads to a system

$$\mathbf{y}_{\text{app}}(\bar{t} + h, \tilde{\omega}) - \mathbf{x}_{\text{app}}(\bar{t}, \tilde{\omega}) + \frac{h}{2} \Bigg(\mathbf{Pr}(\mathbf{x}_{\text{app}}(\bar{t}, \tilde{\omega})) \nabla f(\mathbf{x}_{\text{app}}(\bar{t}, \tilde{\omega}))$$

$$+ \mathbf{Pr}(\mathbf{y}_{\text{app}}(\bar{t} + h, \tilde{\omega})) \nabla f(\mathbf{y}_{\text{app}}(\bar{t} + h, \tilde{\omega})) \Bigg)$$

$$- \frac{\epsilon}{2} \Bigg(\mathbf{Pr}(\mathbf{x}_{\text{app}}(\bar{t}, \tilde{\omega})) + \mathbf{Pr}(\mathbf{y}_{\text{app}}(\bar{t} + h, \tilde{\omega})) \Bigg) \cdot$$

$$\cdot (\mathbf{B}_{\bar{t}+h}(\tilde{\omega}) - \mathbf{B}_{\bar{t}}(\tilde{\omega}))$$

$$= 0$$

of nonlinear equations in $\mathbf{y}_{\text{app}}(\bar{t} + h, \tilde{\omega})$ in general. Linearization at $\mathbf{x}_{\text{app}}(\bar{t}, \tilde{\omega})$ yields a system of linear equations with the predictor $\mathbf{y}_{\text{app}}(\bar{t} + h, \tilde{\omega})$ as solution. The corrector step is defined by the computation of a zero of

$$c : \mathbb{R}^m \to \mathbb{R}^m, \quad \boldsymbol{\alpha} \mapsto \begin{pmatrix} h_1 \left(\mathbf{y}_{\text{app}}(\bar{t} + h, \tilde{\omega}) + \sum_{i=1}^{m} \alpha_i \nabla h_i(\mathbf{x}_{\text{app}}(\bar{t}, \tilde{\omega})) \right) \\ \vdots \\ h_m \left(\mathbf{y}_{\text{app}}(\bar{t} + h, \tilde{\omega}) + \sum_{i=1}^{m} \alpha_i \nabla h_i(\mathbf{x}_{\text{app}}(\bar{t}, \tilde{\omega})) \right) \end{pmatrix},$$

which again can be done by linearization techniques.

The idea of step size control is exactly like for the unconstrained case. See Fig. 5.7 for illustration.

The point $\tilde{\mathbf{x}}^1(\bar{t} + h)$ is accepted as a numerical approximation of $\mathbf{X}_{\text{pr}}(\bar{t} + h, \tilde{\omega})$ if

$$d_M \left(\tilde{\mathbf{x}}^1(\bar{t} + h), \tilde{\mathbf{x}}^2(\bar{t} + h) \right) < \delta,$$

which is often replaced by

$$\| \tilde{\mathbf{x}}^1(\bar{t} + h) - \tilde{\mathbf{x}}^2(\bar{t} + h) \|_2 < \delta$$

for practical reasons.

The first of three problems, which we take from [Stö00], is of low dimension and serves as an example for the visualization of the numerical results.

Problem 12. $n = 3$

$$\text{globmin}_{\mathbf{x}} \left\{ f : \mathbb{R}^3 \to \mathbb{R}, \ (x_1, x_2, x_3)^\top \mapsto x_1^2 + x_2^2; \ 1 - (x_1 - 2)^2 - \frac{x_2^2}{9} + x_3^2 = 0 \right\},$$

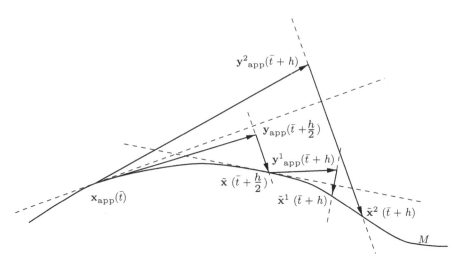

Fig. 5.7 Step size control

This problem has three isolated minimum points:

$(0, 0, \sqrt{3})^\top$ global minimum point, $f\left((0, 0, \sqrt{3})\right) = 0,$

$(0, 0, -\sqrt{3})^\top$ global minimum point, $f\left((0, 0, -\sqrt{3})\right) = 0,$

$(3, 0, 0)^\top$ local minimum point, $f\left((3, 0, 0)\right) = 9.$

If we interpret the variable x_3 as slack variable (the idea of replacing inequalities by equations with squared variables is published in [McShane73]), this constrained global optimization problem in three dimensions with one equality constraint results from the constrained global optimization problem.

Problem 12'. $n = 2$

$$\text{globmin}_{\mathbf{x}} \left\{ f' : \mathbb{R}^2 \to \mathbb{R},\ (x_1, x_2)^\top \mapsto x_1^2 + x_2^2;\ 1 - (x_1 - 2)^2 - \frac{x_2^2}{9} \leq 0 \right\}$$

with one inequality constraint. Problem 12' has two isolated minimum points:

$(0, 0)^\top$ global minimum point, $f'\left((0, 0)\right) = 0,$

$(3, 0)^\top$ local minimum point, $f'\left((3, 0)\right) = 9.$

The inequality constraint

$$1 - (x_1 - 2)^2 - \frac{x_2^2}{9} \leq 0$$

excludes an ellipse from the $\mathbb{R}^2$-plane (Fig. 5.8).

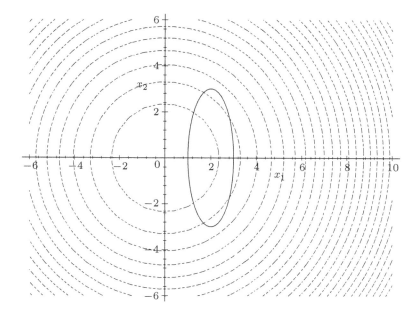

Fig. 5.8 Contour lines of f' and the excluded ellipse

The numerical treatment of Problem 12 with
$$\epsilon = 2.0, \quad \delta = 0.1$$
chosen number of iterations: 200

starting point $\mathbf{x}_0 = \left(10, 1, \sqrt{63 + \frac{1}{9}}\right)^\top$

leads to results in $\mathbb{R}^2$ shown in Fig. 5.9, neglecting the slack variable x_3.

After finding the local minimum point $(3,0)^\top$, the computed path goes on and passes through the global minimum point $(0,0)^\top$. A second run of the same algorithm with the same parameters (hence, only another path is chosen) yields the result shown in Fig. 5.10.

The same type of path is obtained using $\epsilon = 5$ instead of $\epsilon = 2$ (see Fig. 5.11).

If $\epsilon = 1$, then 200 iterations are not enough to leave the local minimum point as shown in Fig. 5.12.

If ϵ is too large, then purely random search dominates (see Fig. 5.13 with $\epsilon = 20$).

Again, the optimal choice of ϵ depends on scaling properties of the functions $f, h_1, \ldots, h_{m+k}$. In the next section, we will come back to this example.

Problem 13. $n = 20$

$$\text{globmin}_{\mathbf{x}} \left\{ f : \mathbb{R}^{20} \to \mathbb{R}, \ \mathbf{x} \mapsto 5(1 - x_1) - \exp\left(-100(x_1 - 0.7)^2\right) + \exp(-9); \right.$$

$$\left. \sum_{i=1}^{20} x_i^2 - 1 = 0 \right\}.$$

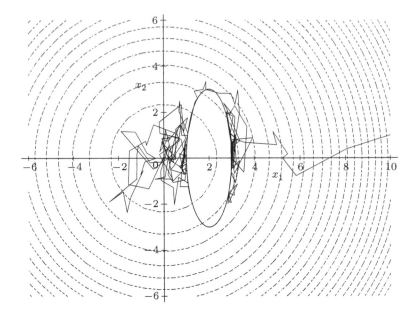

Fig. 5.9 Solution of Problem 12' by solving Problem 12, first path

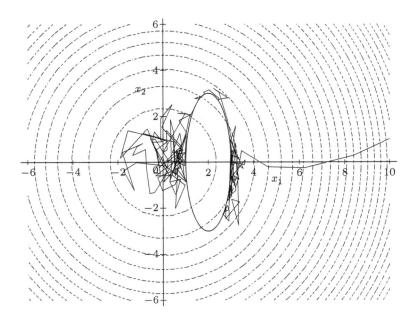

Fig. 5.10 Solution of Problem 12' by solving Problem 12, second path

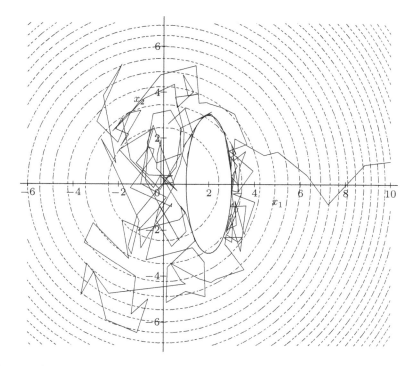

Fig. 5.11 Solution of Problem 12' by solving Problem 12 with a larger ϵ

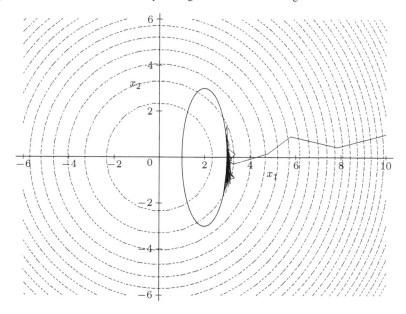

Fig. 5.12 Solution of Problem 12' by solving Problem 12 with a too small ϵ

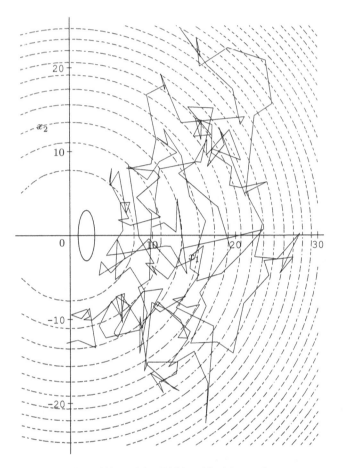

Fig. 5.13 Solution of Problem 12' by solving Problem 12 with a too large ϵ

This problem has a unique global minimum point

$$\mathbf{x}_{gl} = (1, 0, \ldots, 0)^{\top}.$$

Let x_{loc} be the unique local (not global) minimum point of the function (Fig. 5.14)

$$f_1 : \mathbb{R} \to \mathbb{R}, \quad x \mapsto 5(1 - x) - \exp\left(-100(x - 0.7)^2\right) + \exp(-9),$$

then the local minimum points of Problem 13 are given by the set

$$\left\{ \mathbf{x} \in \mathbb{R}^{20}; \, x_1 = x_{loc} \text{ and } \sum_{i=2}^{20} x_i^2 = 1 - x_{loc}^2 \right\},$$

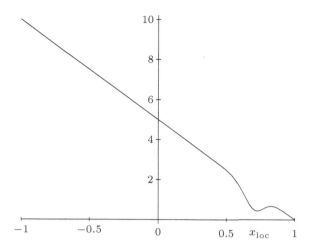

Fig. 5.14 The graph of f_1 for $x \in [-1, 1]$

which is a 18-dimensional sphere with radius $\sqrt{1 - x_{\text{loc}}^2}$ within the affine hyperplane $x_1 = x_{\text{loc}}$. Starting from $(-1, 0, \ldots, 0)^\top$ with $\epsilon = 0.2$, $\delta = 0.1$, and computing 500 iterations, the function value of the computed starting point for the local minimization is equal to $4.67 \cdot 10^{-1}$. Local minimization leads to the global minimum point.

Problem 14. $n = 100$

$$\text{globmin}_{\mathbf{x}} \left\{ f : \mathbb{R}^{100} \to \mathbb{R},\ \mathbf{x} \mapsto 4 + 8x_1^2 - 4\cos(8x_1) + x_3^2 + x_4^2 \right.$$

$$+ (x_5 - 1)^2 + \sum_{i=6}^{100} x_i^2;$$

$$x_4^2 - \sum_{i=5}^{100} x_i^2 + 1 = 0$$

$$\left. x_1^2 - x_2^2 - x_3^2 + 1 = 0 \right\}.$$

This problem has two isolated global minimum points with function value equal to zero and apart from them, at least eight isolated local minimum points. Using a starting point with function value equal to $2.27 \cdot 10^4$ and computing 200 iterations with $\epsilon = 2.0$ and with $\delta = 0.1$, the function value of the computed starting point for the local minimization is equal to $1.30 \cdot 10^2$. Local minimization leads to a global minimum point.

5.4   General Case

Let us consider the constrained global minimization problem

$$\text{globmin}_{\mathbf{x}}\{f(\mathbf{x}); \ h_i(\mathbf{x}) = 0, \quad i = 1,\ldots,m$$

$$h_i(\mathbf{x}) \leq 0, \quad i = m+1,\ldots,m+k\},$$

$$f, h_i : \mathbb{R}^n \to \mathbb{R}, \quad n \in \mathbb{N}, \ m,k \in \mathbb{N}_0,$$

$$f, h_i \in C^2(\mathbb{R}^n, \mathbb{R}), \quad i = 1,\ldots,m+k.$$

Let

$$\mathbf{x}^* \in R = \{\mathbf{x} \in \mathbb{R}^n; \ h_i(\mathbf{x}) = 0, \quad i = 1,\ldots,m$$

$$h_i(\mathbf{x}) \leq 0, \quad i = m+1,\ldots,m+k\}$$

and let $J_{\mathbf{x}^*} = \{j_1, \ldots j_p\} \subseteq \{m+1,\ldots,m+k\}$ be the set of indices j for which

$$h_j(\mathbf{x}^*) = 0.$$

We assume that the gradient vectors of all active constraints in $\mathbf{x}^*$

$$\nabla h_1(\mathbf{x}^*), \ldots, \nabla h_m(\mathbf{x}^*), \nabla h_{j_1}(\mathbf{x}^*), \ldots, \nabla h_{j_p}(\mathbf{x}^*)$$

are linearly independent for each $\mathbf{x}^* \in R$.

Motivated by the numerical analysis of Problems 12 and 12', it seems interesting to transform this constrained global minimization problem to

$$\text{globmin}_{\mathbf{x},\mathbf{s}}\left\{\tilde{f} : \mathbb{R}^{n+k} \to \mathbb{R}, \quad (\mathbf{x},\mathbf{s})^\top \mapsto f(\mathbf{x}); \right.$$

$$h_i(\mathbf{x}) = 0, \quad i = 1,\ldots,m$$

$$\left. h_i(\mathbf{x}) + s_{i-m+n}^2 = 0, \quad i = m+1,\ldots,m+k\right\}.$$

For this approach, we note:

- The number of variables is enlarged by the number of inequality constraints.
- Each minimum point of f with l inactive inequality constraints creates 2^l minimum points of $\tilde{f}$ with the same function value.
- Using the projected curve of steepest descent

$$(\dot{\mathbf{x}},\dot{\mathbf{s}})_{\text{pr}}^\top(t) = -\mathbf{Pr}((\mathbf{x},\mathbf{s})_{\text{pr}}(t))\nabla \tilde{f}((\mathbf{x},\mathbf{s})_{\text{pr}}(t)), \quad (\mathbf{x},\mathbf{s})^\top(0) = (\mathbf{x}_0,\mathbf{s}_0)^\top$$

for the local optimization problem

$$\text{locmin}_{\mathbf{x,s}} \left\{ \tilde{f} : \mathbb{R}^{n+k} \to \mathbb{R}, \quad (\mathbf{x,s})^\top \mapsto f(\mathbf{x}); \right.$$

$$h_i(\mathbf{x}) = 0, \quad i = 1, \ldots, m$$

$$\left. h_i(\mathbf{x}) + s_{i-m+n}^2 = 0, \quad i = m+1, \ldots, m+k \right\},$$

all active inequality constraints at the starting point $\mathbf{x}_0$ remain active and all nonactive inequality constraints at the starting point $\mathbf{x}_0$ remain nonactive along the projected curve of steepest descent. Hence, this idea does not work for local minimization.

Fortunately, the third statement is no longer true when we use the randomized projected curve of steepest descent for global minimization as can be observed in Figs. 5.9–5.11. The reason is that by the additive term $\epsilon \mathbf{S}_t(\omega)$ in the Fisk–Stratonovich integral equation, the randomized projected curve of steepest descent is no longer determined by an autonomous system. Therefore, it is theoretically feasible to consider only constrained global minimization problems with equality constraints using slack variables. But if there is a huge number of inequality constraints, then the following active set method may be a better alternative.

Choose a point

$$\mathbf{x}_0 \in R = \{\mathbf{x} \in \mathbb{R}^n; \ h_i(\mathbf{x}) = 0, \quad i = 1, \ldots, m$$

$$h_i(\mathbf{x}) \leq 0, \quad i = m+1, \ldots, m+k\}$$

and let $J_{\mathbf{x}_0} \subseteq \{m+1, \ldots, m+k\}$ be the set of indices j for which

$$h_j(\mathbf{x}_0) = 0,$$

then we consider the constrained global minimization problem in $n + |J_{\mathbf{x}_0}|$ variables

$$\text{globmin}_{(\mathbf{x,s})} \left\{ f_{\mathbf{x}_0} : \mathbb{R}^{n+|J_{\mathbf{x}_0}|} \to \mathbb{R}, \quad (\mathbf{x,s})^\top \mapsto f(\mathbf{x}); \right.$$

$$h_i(\mathbf{x}) = 0, \quad i = 1, \ldots, m$$

$$\left. h_j(\mathbf{x}) + s_{n-m+j}^2 = 0, j \in J_{\mathbf{x}_0} \right\}.$$

Using the randomized projected curve of steepest descent for this constrained global minimization problem with $m + |J_{\mathbf{x}_0}|$ equality constraints and with starting point $(\mathbf{x}_0, \mathbf{0})^\top$, we perform only one predictor and corrector step with step size control. In the special case of no equality constraints, i.e., $m = 0$, and of no active inequality constraints at $\mathbf{x}_0$, this means to perform a single iteration, including step size control, for an unconstrained problem. The step size control has to be modified such that the first n components of the computed point $(\mathbf{x}(t_1, \tilde{\omega}), \mathbf{s}_{\tilde{\omega}}^1)^\top$ (i.e., $\mathbf{x}(t_1, \tilde{\omega})$) are in the feasible region R. Then, we consider the constrained global minimization problem in $n + |J_{\mathbf{x}(t_1, \tilde{\omega})}|$ variables

$$\operatorname*{globmin}_{(\mathbf{x}, \mathbf{s})}\left\{ f_{\mathbf{x}(t_1, \tilde{\omega})} : \mathbb{R}^{n + |J_{\mathbf{x}(t_1, \tilde{\omega})}|} \to \mathbb{R}, \quad (\mathbf{x}, \mathbf{s})^\top \mapsto f(\mathbf{x}); \right.$$

$$h_i(\mathbf{x}) = 0, \quad i = 1, \ldots, m$$

$$\left. h_j(\mathbf{x}) + s_{n-m+j}^2 = 0, \quad j \in J_{\mathbf{x}(t_1, \tilde{\omega})} \right\},$$

where $J_{\mathbf{x}(t_1, \tilde{\omega})} \subseteq \{m+1, \ldots, m+k\}$ is the set of indices j for which

$$h_j(\mathbf{x}(t_1, \tilde{\omega})) = 0.$$

It may be that new inequality constraints are now active at $\mathbf{x}(t_1, \tilde{\omega})$ and, since we used a randomization, it may be that active constraints became nonactive. Now, we handle this constrained global minimization problem with $m + |J_{\mathbf{x}(t_1, \tilde{\omega})}|$ equality constraints in the same way as before using $(\mathbf{x}(t_1, \tilde{\omega}), \mathbf{0})^\top$ as a starting point. Consequently, we compute only one point $(\mathbf{x}(t_2, \tilde{\omega}), \mathbf{s}_{\tilde{\omega}}^2)^\top$ and consider the next constrained global minimization problem in $n + |J_{\mathbf{x}(t_2, \tilde{\omega})}|$ variables

$$\operatorname*{globmin}_{(\mathbf{x}, \mathbf{s})}\left\{ f_{\mathbf{x}(t_2, \tilde{\omega})} : \mathbb{R}^{n + |J_{\mathbf{x}(t_2, \tilde{\omega})}|} \to \mathbb{R}, \quad (\mathbf{x}, \mathbf{s})^\top \mapsto f(\mathbf{x}); \right.$$

$$h_i(\mathbf{x}) = 0, \quad i = 1, \ldots, m$$

$$\left. h_j(\mathbf{x}) + s_{n-m+j}^2 = 0, \quad j \in J_{\mathbf{x}(t_2, \tilde{\omega})} \right\},$$

with starting point $(\mathbf{x}(t_2, \tilde{\omega}), \mathbf{0})^\top$, where $J_{\mathbf{x}(t_2, \tilde{\omega})} \subseteq \{m+1, \ldots, m+k\}$ is the set of indices j for which

$$h_j(\mathbf{x}(t_2, \tilde{\omega})) = 0$$

and so on.

We demonstrate the performance of this method by two test problems.

Problem 15. $n = 2$

$$\operatorname*{globmin}_{\mathbf{x}}\left\{ f : \mathbb{R}^2 \to \mathbb{R}, \, \mathbf{x} \mapsto 6x_1^2 - \cos(12x_1) + 6x_2^2 - \cos(12x_2) + 2; \right.$$

$$0.1 \le x_1 \le 1;$$

$$\left. 0.1 \le x_2 \le 1 \right\}.$$

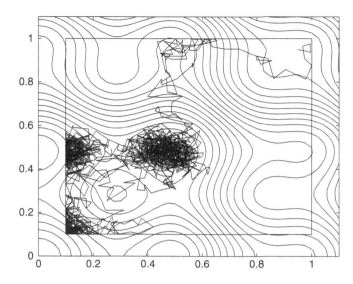

Fig. 5.15 Problem 15, contour lines, $\epsilon = 1$, 1, 500 points

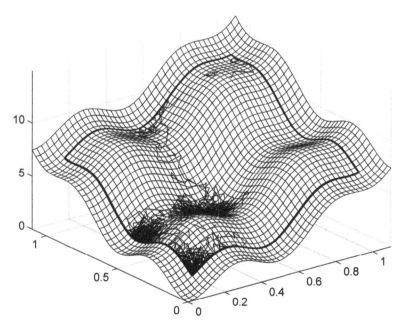

Fig. 5.16 Problem 15, function values, $\epsilon = 1$, 1, 500 points

This problem has a unique isolated global minimum point $(0.1, 0.1)^\top$. Using $(0.95, 0.95)^\top$ as a starting point and $\epsilon = 1$, the results are shown in Figs. 5.15 and 5.16:

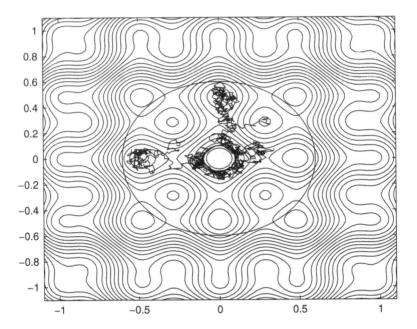

Fig. 5.17 Problem 16, contour lines, $\epsilon = 1, 1, 500$ points

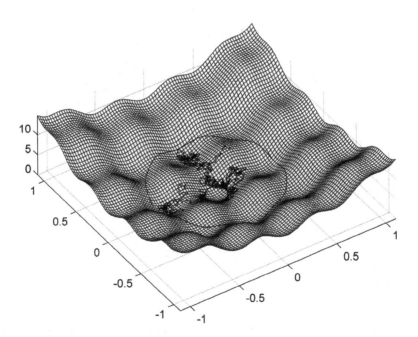

Fig. 5.18 Problem 16, function values, $\epsilon = 1, 1, 500$ points

Problem 16. $n = 2$

$$\text{globmin}_{\mathbf{x}} \left\{ f : \mathbb{R}^2 \to \mathbb{R},\ \mathbf{x} \mapsto 6x_1^2 - \cos(12x_1) + 6x_2^2 - \cos(12x_2) + 2; \right.$$

$$\left. 0.1 \leq x_1^2 + x_2^2 \leq 1 \right\}.$$

This problem has exactly four isolated global minimum points

$$(-0.1, 0)^\top, (0, 0.1)^\top, (0.1, 0)^\top, (0, -0.1)^\top.$$

Using $(0.3, 0.3)^\top$ as a starting point and $\epsilon = 1$, we obtain results summarized in Figs. 5.17 and 5.18.

Chapter 6
Vector Optimization

6.1 Introduction

The stochastic approach for global optimization problems investigated in the last chapters can be used for vector optimization problems also. For that, we introduce a partial ordering $\leq_P$ on $\mathbb{R}^n$ by

$$\mathbf{u} \leq_P \mathbf{v} \quad :\Longleftrightarrow \quad u_i \leq v_i \quad i = 1, \ldots, n, \quad \mathbf{u}, \mathbf{v} \in \mathbb{R}^n$$

and consider constrained global vector minimization problems of the following type:

$$\text{vglobmin}_{\mathbf{x}} \{\mathbf{f}(\mathbf{x}); \; h_i(\mathbf{x}) = 0, \quad i = 1, \ldots, m$$

$$h_i(\mathbf{x}) \leq 0, \quad i = m + 1, \ldots, m + k\},$$

$$\mathbf{f} = \begin{pmatrix} f_1 \\ \vdots \\ f_l \end{pmatrix} : \mathbb{R}^n \to \mathbb{R}^l,$$

$$h_i : \mathbb{R}^n \to \mathbb{R}, \quad f_j : \mathbb{R}^n \to \mathbb{R}$$

$$n, l \in \mathbb{N}, \quad m, k, \in \mathbb{N}_0,$$

$$f_j, h_i \in C^2(\mathbb{R}^n, \mathbb{R}), \quad i = 1, \ldots, m + k, \quad j = 1, \ldots, l.$$

This means that we have to compute at least one point

$$\mathbf{x}_{gl} \in R := \{\mathbf{x} \in \mathbb{R}^n; \; h_i(\mathbf{x}) = 0, \quad i = 1, \ldots, m$$

$$h_i(\mathbf{x}) \leq 0, \quad i = m + 1, \ldots, m + k\} \neq \emptyset$$

S. Schäffler, *Global Optimization: A Stochastic Approach*, Springer Series in Operations Research and Financial Engineering, DOI 10.1007/978-1-4614-3927-1_6, © Springer Science+Business Media New York 2012

such that there exists no point $\mathbf{x} \in R$ with

$$\mathbf{f}(\mathbf{x}_{gl}) \neq \mathbf{f}(\mathbf{x}) \quad \text{and} \quad \mathbf{f}(\mathbf{x}) \leq_P \mathbf{f}(\mathbf{x}_{gl}).$$

A point $\mathbf{x}_{gl}$ of this type is called a global Pareto minimum point. Since global Pareto minimum points are not unique in general (because $\leq_P$ is just a partial order), one is interested in computing as much global Pareto minimum points as possible. A point $\mathbf{x}_1 \in R$ is said to be dominated by a point $\mathbf{x}_2 \in R$ if

$$\mathbf{f}(\mathbf{x}_1) \neq \mathbf{f}(\mathbf{x}_2) \quad \text{and} \quad \mathbf{f}(\mathbf{x}_2) \leq_P \mathbf{f}(\mathbf{x}_1) \quad (\text{i.e.}, \mathbf{f}(\mathbf{x}_2) <_P \mathbf{f}(\mathbf{x}_1)).$$

The book by Johannes Jahn [Jahn10] presents fundamentals and important results of vector optimization in a general setting.

6.2   The Curve of Dominated Points

We investigate unconstrained local vector minimization problems:

$$\operatorname*{vlocmin}_{\mathbf{x}} \{\mathbf{f}(\mathbf{x})\}, \quad \mathbf{f} = \begin{pmatrix} f_1 \\ \vdots \\ f_l \end{pmatrix} : \mathbb{R}^n \to \mathbb{R}^l, \quad n, l \in \mathbb{N},$$

$$f_j \in C^2(\mathbb{R}^n, \mathbb{R}), \quad j = 1, \ldots, l.$$

Hence, we have to compute at least one point $\mathbf{x}_{loc} \in \mathbb{R}^n$ such that there exists no point $\mathbf{x} \in U(\mathbf{x}_{loc})$ with

$$\mathbf{f}(\mathbf{x}_{loc}) \neq \mathbf{f}(\mathbf{x}) \quad \text{and} \quad \mathbf{f}(\mathbf{x}) \leq_P \mathbf{f}(\mathbf{x}_{loc}),$$

where $U(\mathbf{x}_{loc}) \subseteq \mathbb{R}^n$ is an open neighborhood of $\mathbf{x}_{loc}$.

We are going to generalize the curve of steepest descent for vector minimization problems. For this purpose, for $\mathbf{x} \in \mathbb{R}^n$, we consider the convex hull $C_{con}(\mathbf{x})$ of

$$\nabla f_1(\mathbf{x}), \ldots, \nabla f_l(\mathbf{x})$$

and compute the unique vector $\mathbf{v}(\mathbf{x}) \in C_{con}(\mathbf{x})$ with

$$\|\mathbf{v}(\mathbf{x})\|_2 \leq \|\mathbf{w}(\mathbf{x})\|_2, \quad \text{for all } \mathbf{w}(\mathbf{x}) \in C_{con}(\mathbf{x})$$

by solving the following global quadratic minimization problem:

$$\text{globmin}_{\mathbf{z}}\left\{\left\|\sum_{i=1}^{l} z_i \nabla f_i(\mathbf{x})\right\|_2^2 ; \quad \sum_{i=1}^{l} z_i - 1 = 0 \right.$$

$$\left. -z_i \leq 0 \quad i = 1, \ldots, l\right\}, \quad \mathbf{x} \in \mathbb{R}^n.$$

With a global minimum point $\mathbf{z}_{\text{gl}}(\mathbf{x})$, we obtain $\mathbf{v}(\mathbf{x}) \in C_{\text{con}}(\mathbf{x})$ by the following function:

$$\mathbf{v} : \mathbb{R}^n \to \mathbb{R}^n, \quad \mathbf{x} \mapsto \sum_{i=1}^{l} \mathbf{z}_{\text{gl}}(\mathbf{x})_i \nabla f_i(\mathbf{x}).$$

It is proven in [Schä.etal02] that

- Either $\mathbf{v}(\mathbf{x}) = \mathbf{0}$ or $-\mathbf{v}(\mathbf{x})$ is a descent direction for all $f_i, i = 1, \ldots, l$.
- The function $\mathbf{v}$ is locally Lipschitz continuous.

Using the function $\mathbf{v}$, we have found an analogon of the gradient in real-valued optimization.

The curve of dominated points

$$\mathbf{x} : \mathbb{D} \subseteq [0, \infty) \to \mathbb{R}^n$$

with starting point $\mathbf{x}_0 \in \mathbb{R}^n$ is defined by the solution of the initial value problem

$$\dot{\mathbf{x}}(t) = -\mathbf{v}(\mathbf{x}(t)), \quad \mathbf{x}(0) = \mathbf{x}_0.$$

In the following theorem, we summarize some important properties of the above initial value problem.

Theorem 6.1 *Consider*

$$\mathbf{f} = \begin{pmatrix} f_1 \\ \vdots \\ f_l \end{pmatrix} : \mathbb{R}^n \to \mathbb{R}^l, \quad n, l \in \mathbb{N}, \quad f_j \in C^2(\mathbb{R}^n, \mathbb{R}), \quad j = 1, \ldots, l,$$

and let the set

$$L_{\mathbf{f}, \mathbf{x}_0} := \{\mathbf{x} \in \mathbb{R}^n; \mathbf{f}(\mathbf{x}) \leq_P \mathbf{f}(\mathbf{x}_0)\}$$

be bounded, then we obtain:

(i) The initial value problem

$$\dot{\mathbf{x}}(t) = -\mathbf{v}(\mathbf{x}(t)), \quad \mathbf{x}(0) = \mathbf{x}_0,$$

has a unique solution $\mathbf{x} : [0, \infty) \to \mathbb{R}^n$.

(ii) *Either*

$$\mathbf{x} \equiv \mathbf{x}_0 \quad \textit{iff} \quad \mathbf{v}(\mathbf{x}_0) = \mathbf{0}$$

or

$$\mathbf{f}(\mathbf{x}(t+h)) <_P \mathbf{f}(\mathbf{x}(t)) \quad \textit{for all} \quad t, h \in [0, \infty), \, h > 0.$$

(iii) *For each*

$$\tilde{\mathbf{z}} \in \left\{ \mathbf{z} \in \mathbb{R}^l; \sum_{i=1}^{l} z_i = 1, \quad z_i \geq 0, \, i = 1, \dots, l \right\},$$

there exists a point $\mathbf{x}_{stat} \in \mathbb{R}^n$ *with*

$$\lim_{t \to \infty} f_{\tilde{\mathbf{z}}}(\mathbf{x}(t)) = f_{\tilde{\mathbf{z}}}(\mathbf{x}_{stat}) \quad \textit{and} \quad \mathbf{v}(\mathbf{x}_{stat}) = \mathbf{0},$$

where

$$f_{\tilde{\mathbf{z}}} : \mathbb{R}^n \to \mathbb{R}, \quad \mathbf{x} \mapsto \sum_{i=1}^{l} \tilde{z}_i f_i(\mathbf{x}).$$

The proof uses exactly the same ideas as the one of Theorem 2.1.

For the numerical approximation of the curve of dominated points, only function values of $\mathbf{v}$ are available, since $\mathbf{v}$ is not differentiable in general.

Example 6.2. Let us investigate a local vector minimization problem with

$$\mathbf{f} : \mathbb{R}^2 \to \mathbb{R}^3, \quad \mathbf{x} \mapsto \begin{pmatrix} f_1(\mathbf{x}) \\ f_2(\mathbf{x}) \\ f_3(\mathbf{x}) \end{pmatrix} = \begin{pmatrix} x_1 + x_2 \\ \sqrt{0.06(x_2 - x_1)^2 - \cos(1.2(x_2 - x_1)) + 2)} - 1 \\ \arctan(x_2 - x_1) \end{pmatrix}.$$

The set of all Pareto minimum points is given by

$$\left\{ \mathbf{x} \in \mathbb{R}^2; \, \mathbf{x} \text{ is a Pareto minimum point of } \mathbf{g} : \mathbb{R}^2 \to \mathbb{R}^2, \quad \mathbf{x} \mapsto \begin{pmatrix} f_2(\mathbf{x}) \\ f_3(\mathbf{x}) \end{pmatrix} \right\}$$

and their function values are visualized in Fig. 6.1.

Figure 6.2 shows function values of the curve of dominated points according to $\mathbf{f}$ with starting point $(-1, 1)^{\top}$.

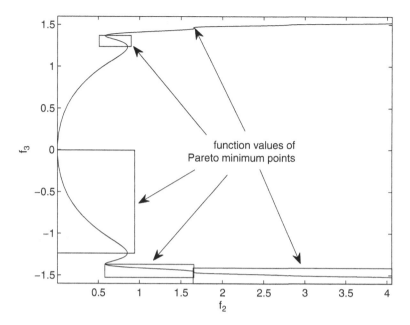

Fig. 6.1 Function values of f_2 and f_3

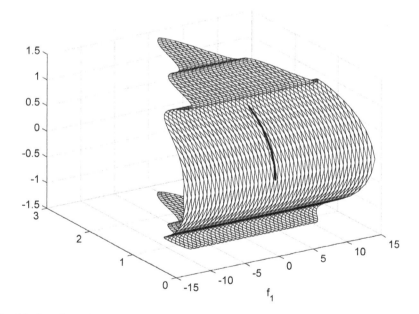

Fig. 6.2 Function values of the curve of dominated points according to **f**

6.3   A Randomized Curve of Dominated Points

Now, we investigate unconstrained global vector minimization problems

$$
\text{vglobmin}_{\mathbf{x}}\{\mathbf{f}(\mathbf{x})\}, \quad \mathbf{f} = \begin{pmatrix} f_1 \\ \vdots \\ f_l \end{pmatrix} : \mathbb{R}^n \to \mathbb{R}^l, \quad n, l \in \mathbb{N},
$$

$$
f_j \in C^2(\mathbb{R}^n, \mathbb{R}), \quad j = 1, \dots, l.
$$

With the Wiener space $(\Omega, \mathcal{B}(\Omega), W)$ and the Brownian Motion $\{\mathbf{B}_t\}_{t \in [0,\infty)}$, we are interested in the investigation of a stochastic process $\{\mathbf{X}_t\}_{t \in [0,\infty)}$ with

$$
\mathbf{X}_t : \Omega \to \mathbb{R}^n, \quad t \in [0, \infty)
$$

given by

$$
\mathbf{X}_t(\omega) = \mathbf{x}_0 - \int_0^t \mathbf{v}(\mathbf{X}_\tau(\omega)) d\tau + \epsilon \left(\mathbf{B}_t(\omega) - \mathbf{B}_0(\omega) \right), \quad \omega \in \Omega,
$$

where $\mathbf{x}_0 \in \mathbb{R}^n$ and $\epsilon > 0$ are fixed. This equation can be interpreted as a randomized curve of dominated points. In order to analyze the randomized curve of dominated points, it is necessary to formulate assumptions on the function $\mathbf{v}$.

Assumption 6.2 *There exists a real number $\epsilon > 0$ such that*

$$
\mathbf{x}^\top \mathbf{v}(\mathbf{x}) \geq \frac{1 + n\epsilon^2}{2} \max\{1, \|\mathbf{v}(\mathbf{x})\|_2\}
$$

for all $\mathbf{x} \in \{\mathbf{w} \in \mathbb{R}^n; \|\mathbf{w}\|_2 > \rho\}$ for some $\rho \in \mathbb{R}, \rho > 0$.

In the following theorem, we study properties of the randomized curve of dominated points analogously to Theorems 3.3 and 3.6, respectively.

Theorem 6.3 *Consider the unconstrained global vector minimization problem*

$$
\text{vglobmin}_{\mathbf{x}}\{\mathbf{f}(\mathbf{x})\}, \quad \mathbf{f} = \begin{pmatrix} f_1 \\ \vdots \\ f_l \end{pmatrix} : \mathbb{R}^n \to \mathbb{R}^l, \quad n, l \in \mathbb{N},
$$

$$
f_j \in C^2(\mathbb{R}^n, \mathbb{R}), \quad j = 1, \dots, l.
$$

and let the following Assumption 6.2 be fulfilled:

There exists a real number $\epsilon > 0$ such that

$$\mathbf{x}^\top \mathbf{v}(\mathbf{x}) \geq \frac{1 + n\epsilon^2}{2} \max\{1, \|\mathbf{v}(\mathbf{x})\|_2\}$$

for all $\mathbf{x} \in \{\mathbf{w} \in \mathbb{R}^n; \|\mathbf{w}\|_2 > \rho\}$ *for some* $\rho \in \mathbb{R}$, $\rho > 0$,

then we obtain:

(i) *Using the Wiener space* $(\Omega, \mathcal{B}(\Omega), W)$ *and* $\epsilon > 0$ *from Assumption 6.2, the integral equation*

$$\mathbf{X}(t, \omega) = \mathbf{x}_0 - \int_0^t \mathbf{v}(\mathbf{X}(\tau, \omega)) d\tau + \epsilon \left(\mathbf{B}_t(\omega) - \mathbf{B}_0(\omega) \right), \quad t \in [0, \infty), \quad \omega \in \Omega$$

has a unique solution $\mathbf{X} : [0, \infty) \times \Omega \to \mathbb{R}^n$ *for each* $\mathbf{x}_0 \in \mathbb{R}^n$.
(ii) *For each* $t \in [0, \infty)$, *the mapping*

$$\mathbf{X}_t : \Omega \to \mathbb{R}^n, \quad \omega \mapsto \mathbf{X}(t, \omega)$$

is an n-dimensional random variable (therefore $\mathcal{B}(\Omega) - \mathcal{B}(\mathbb{R}^n)$ *measurable), and its probability distribution is given by a Lebesgue density function*

$$p_t : \mathbb{R}^n \to \mathbb{R}$$

with

$$\lim_{t \to \infty} p_t(\mathbf{x}) = p(\mathbf{x}) \quad \text{for all} \quad \mathbf{x} \in \mathbb{R}^n,$$

where $p : \mathbb{R}^n \to \mathbb{R}$ *is a Lebesgue density function of a random variable* $X : \Omega \to \mathbb{R}^n$.
(iii) *Choose any* $r > 0$ *and let* $\mathbf{x}_{gl}$ *be any global Pareto minimum point of the unconstrained vector minimization problem. Using the stopping time*

$$st : \Omega \to \mathbb{R} \cup \{\infty\},$$

$$\omega \mapsto \begin{cases} \inf \left\{ t \geq 0; \|\mathbf{X}(t, \omega) - \mathbf{x}_{gl}\|_2 \leq r \right\} & \text{if } \left\{ t \geq 0; \|\mathbf{X}(t, \omega) - \mathbf{x}_{gl}\|_2 \leq r \right\} \neq \emptyset \\ \\ \infty & \text{if } \left\{ t \geq 0; \|\mathbf{X}(t, \omega) - \mathbf{x}_{gl}\|_2 \leq r \right\} = \emptyset \end{cases},$$

we obtain:

a. $W(\{\omega \in \Omega; st(\omega) < \infty\}) = 1.$
b. *For the expectation of st holds:*

$$\mathbb{E}(st) < \infty.$$

Again, the ideas used for the proof of this theorem are the same as the ones used for the proofs of Theorems 3.3 and 3.6, respectively.

Since the function **v** cannot be interpreted as a gradient of a given potential function in general, the invariant measure denoted by p in Theorem 6.3 is not known explicitly.

6.4 An Euler Method

Since the function **v** is not differentiable in general, the numerical approximation of the randomized curve of dominated points has to use only function values of **v** as in the following Euler method.

Step 0: (Initialization)
   Choose $x_0 \in \mathbb{R}^n$ and $\epsilon, \delta > 0$,
   Choose maxit $\in \mathbb{N}$,
   $j := 0$,
   goto step 1.

In step 0, the starting point x_0, the parameter ϵ according to Assumption 6.2, the parameter $\delta > 0$ according to the step size control, and the maximal number of iterations have to be determined by the user.

Step 1: (Evaluation of v)
   $h := 1$,
   compute $v(x_j)$
   goto step 2.

The initial value $h_{\max}$ of the step size is chosen equal to 1.

Step 2: (Pseudorandom Numbers)
   Compute $2n$ stochastically independent $\mathcal{N}(0, 1)$ Gaussian distributed
   pseudorandom numbers $p_1, \ldots p_n, q_1, \ldots, q_n \in \mathbb{R}$,
   goto step 3.

In this step, the choice of the path is determined successively by the computer.

Step 3: (Computation of x_{j+1}^2 by one step with step size h)

$$x_{j+1}^2 := x_j - hv(x_j) + \epsilon\sqrt{\frac{h}{2}}\begin{pmatrix} p_1 + q_1 \\ \vdots \\ p_n + q_n \end{pmatrix}.$$

   goto step 4.

x_{j+1}^2 is computed by a step with starting point x_j using the step size h.

Step 4: (Computation of $\mathbf{x}_{\frac{h}{2}}$)

$$\mathbf{x}_{\frac{h}{2}} := \mathbf{x}_j - \frac{h}{2}\mathbf{v}(\mathbf{x}_j) + \epsilon\sqrt{\frac{h}{2}}\begin{pmatrix} p_1 \\ \vdots \\ p_n \end{pmatrix}.$$

goto step 5.

$\mathbf{x}_{\frac{h}{2}}$ is computed by a step with starting point $\mathbf{x}_j$ using the step size $\frac{h}{2}$.

Step 5: (Evaluation of v)
   compute $\mathbf{v}(\mathbf{x}_{\frac{h}{2}})$
   goto step 6.

Step 6: (Computation of $\mathbf{x}^1_{j+1}$ by two steps with step size $\frac{h}{2}$)

$$\mathbf{x}^1_{j+1} := \mathbf{x}_{\frac{h}{2}} - \frac{h}{2}\mathbf{v}(\mathbf{x}_{\frac{h}{2}}) + \epsilon\sqrt{\frac{h}{2}}\begin{pmatrix} q_1 \\ \vdots \\ q_n \end{pmatrix}.$$

goto step 7.

$\mathbf{x}^1_{j+1}$ is computed by a step with starting point $\mathbf{x}_{\frac{h}{2}}$ using the step size $\frac{h}{2}$.

Step 7: (Acceptance condition)
   If $\|\mathbf{x}^1_{j+1} - \mathbf{x}^2_{j+1}\|_2 < \delta$,
      then
                  $\mathbf{x}_{j+1} := \mathbf{x}^1_{j+1}$,
                  print $\left(j + 1, \mathbf{x}_{j+1}, \mathbf{f}\left(\mathbf{x}_{j+1}\right)\right)$,
                  goto step 8.
      else
                  $h := \frac{h}{2}$,
                  goto step 3.

Step 8: (Termination condition)
   If $j + 1 <$ maxit,
      then
                  $j := j + 1$,
                  goto step 1.
      else
                  STOP.

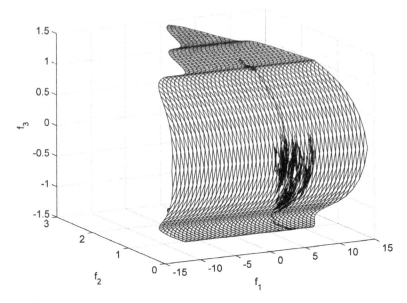

Fig. 6.3 Problem 17, function values of the randomized curve of dominated points, $\epsilon = 0.3$, 1,500 points

It is important to notice that this algorithm should be used to compute as many global Pareto minimum points as possible. An example where this is done successfully is given in [Schä.etal02] with $n = 20$ and $l = 2$.

In order to demonstrate properties of the randomized curve of steepest descent, we come back to Example 6.2.

Problem 17. $n = 2$

$$\operatorname*{vglobmin}_{\mathbf{x}}\left\{\mathbf{f} : \mathbb{R}^2 \to \mathbb{R}^3, \mathbf{x} \mapsto \left(\begin{array}{c} x_1 + x_2 \\ \sqrt{0.06(x_2 - x_1)^2 - \cos(1.2(x_2 - x_1)) + 2)} - 1 \\ \arctan(x_2 - x_1) \end{array}\right)\right\}.$$

Figures 6.3 and 6.4 show function values of the randomized curve of dominated points from two different points of view. It can be noticed that the randomized curve of dominated points behaves like a purely random search near Pareto minimum points.

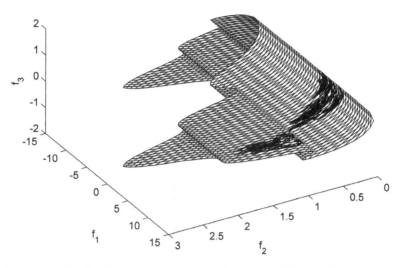

Fig. 6.4 Problem 17, function values of the randomized curve of dominated points, $\epsilon = 0.3$, 1,500 points

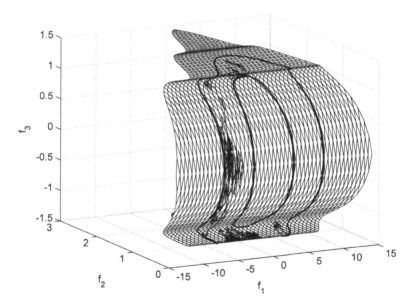

Fig. 6.5 Problem 18, function values, $\epsilon = 0.3$, 5,000 points

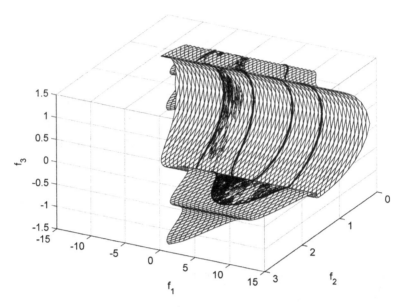

Fig. 6.6 Problem 18, function values, $\epsilon = 0.3$, 5,000 points

6.5 Constraints

Constrained global vector minimization problems

$$
\begin{aligned}
&\underset{\mathbf{x}}{\text{vglobmin}}\{\mathbf{f}(\mathbf{x}); \ h_i(\mathbf{x}) = 0, \quad i = 1, \ldots, m \\
&\qquad\qquad\qquad h_i(\mathbf{x}) \leq 0, \quad i = m+1, \ldots, m+k\}, \\[2mm]
&\mathbf{f} = \begin{pmatrix} f_1 \\ \vdots \\ f_l \end{pmatrix} : \mathbb{R}^n \to \mathbb{R}^l, \\[2mm]
&\qquad\qquad h_i : \mathbb{R}^n \to \mathbb{R}, \quad f_j : \mathbb{R}^n \to \mathbb{R} \\
&\qquad\qquad n, l \in \mathbb{N}, \quad m, k, \in \mathbb{N}_0, \\
&\quad f_j, h_i \in C^2(\mathbb{R}^n, \mathbb{R}), \quad i = 1, \ldots, m+k, \quad j = 1, \ldots, l.
\end{aligned}
$$

can be handled in the same way as in Sect. 5.3 using the randomized projected curve of dominated points for equality constraints.

The penalty approach of Sect. 5.2 can be generalized to vector minimization problems using the functions

$$f_{j,\text{penalty},c_p} : \mathbb{R}^n \to \mathbb{R}, \quad \mathbf{x} \mapsto f_j(\mathbf{x}) + c_p \left(\sum_{i=1}^{m} h_i(\mathbf{x})^4 + \sum_{i=m+1}^{m+k} (P(h_i(\mathbf{x})))^4 \right)$$

with $j = 1, \ldots, l$, and with

$$P : \mathbb{R} \to \mathbb{R}, \quad x \mapsto \begin{cases} x & \text{for} \quad x > 0 \\ 0 & \text{for} \quad x \leq 0 \end{cases},$$

where $\{c_p\}_{p \in \mathbb{N}_0}$ is a sequence of positive, strictly monotonic increasing real numbers with

$$\lim_{p \to \infty} c_p = \infty.$$

In addition, inequalities can be handled analogously to Sect. 5.4, which we demonstrate by means of the following:

Problem 18. $n = 2$

$$\underset{\mathbf{x}}{\text{vglobmin}} \left\{ \mathbf{f} : \mathbb{R}^2 \to \mathbb{R}^3, \mathbf{x} \mapsto \begin{pmatrix} x_1 + x_2 \\ \sqrt{0.06(x_2 - x_1)^2 - \cos(1.2(x_2 - x_1)) + 2)} - 1 \\ \arctan(x_2 - x_1) \end{pmatrix} \right.$$

$$\left. 4 \leq x_1^2 + x_2^2 \leq 25 \right\}.$$

Figures 6.5 and 6.6 show function values from two different points of view.

Appendix A
A Short Course in Probability Theory

Consider a nonempty set Ω. With $\mathcal{P}(\Omega)$, we denote the power set of Ω consisting of all subsets of Ω. Using $\bar{\mathbb{R}} := \mathbb{R} \cup \{\pm\infty\}$, we expand the algebraic structure of $\mathbb{R}$ to $\bar{\mathbb{R}}$ by

$$a + (\pm\infty) = (\pm\infty) + a = (\pm\infty) + (\pm\infty) = (\pm\infty), \quad +\infty - (-\infty) = +\infty,$$

$$a \cdot (\pm\infty) = (\pm\infty) \cdot a = \begin{cases} (\pm\infty), & \text{für } a > 0, \\ 0, & \text{für } a = 0, \\ (\mp\infty), & \text{für } a < 0, \end{cases}$$

$$(\pm\infty) \cdot (\pm\infty) = +\infty, \quad (\pm\infty) \cdot (\mp\infty) = -\infty, \quad \frac{a}{\pm\infty} = 0$$

for all $a \in \mathbb{R}$. With $-\infty < a$ and $a < \infty$ for all $a \in \mathbb{R}$, $(\bar{\mathbb{R}}, \leq)$ remains an ordered set, but not an ordered field.

Let $\mathcal{F}$ be a family of subsets of Ω with $\emptyset \in \mathcal{F}$ then a function

$$\mu : \mathcal{F} \to \bar{\mathbb{R}}$$

is called a *measure* on $\mathcal{F}$ if the following conditions are fulfilled:

(M1) $\mu(A) \geq 0$ *for all* $A \in \mathcal{F}$
(M2) $\mu(\emptyset) = 0$
(M3) *For each sequence* $\{A_i\}_{i \in \mathbb{N}}$ *of pairwise disjoint sets with* $A_i \in \mathcal{F}$
$i \in \mathbb{N}$, *and* $\bigcup_{i=1}^{\infty} A_i \in \mathcal{F}$ *holds:*

$$\mu\left(\bigcup_{i=1}^{\infty} A_i\right) = \sum_{i=1}^{\infty} \mu(A_i) \quad (\sigma\text{-additivity}).$$

S. Schäffler, *Global Optimization: A Stochastic Approach*, Springer Series in Operations Research and Financial Engineering, DOI 10.1007/978-1-4614-3927-1,
© Springer Science+Business Media New York 2012

Let $\{B_i\}_{i \in \mathbb{N}}$ be a sequence with $B_i \subseteq B_{i+1}$, $B_i \in \mathcal{F}$, and $\bigcup_{i=1}^{\infty} B_i = \Omega$. If $\mu(B_i) <$ ∞ for all $i \in \mathbb{N}$, then μ is called σ-*finite*.

If one is only interested in measures on power sets, these measures have limited properties and are not adapted for practical applications. Therefore, we consider the following special class of families of subsets of Ω.

A family of subsets of Ω is called a σ-*field on* Ω if the following conditions are fulfilled:

(S1) $\Omega \in \mathcal{S}$.
(S2)   If $A \in \mathcal{S}$, then $A^c := \Omega \setminus A \in \mathcal{S}$.
(S3)   If $A_i \in \mathcal{S}$, $i \in \mathbb{N}$, then $\bigcup_{i=1}^{\infty} A_i \in \mathcal{S}$.

Let I be any nonempty set and let $\mathcal{S}_i$ be a σ-field on Ω for all $i \in I$; then $\bigcap_{i \in I} \mathcal{S}_i$ is again a σ-field on Ω. Assume $\mathcal{F} \subseteq \mathcal{P}(\Omega)$ and let Σ be the set of all σ-fields on Ω with

$$\mathcal{S} \in \Sigma \quad \Longleftrightarrow \quad \mathcal{F} \subseteq \mathcal{S},$$

then

$$\sigma(\mathcal{F}) := \bigcap_{S \in \Sigma} \mathcal{S}$$

is called σ-*field generated by* $\mathcal{F}$. For $\Omega = \mathbb{R}^n$, $n \in \mathbb{N}$, we consider the σ-field

$$\mathcal{B}^n = \sigma(\{([a_1, b_1) \times \ldots \times [a_n, b_n)) \cap \mathbb{R}^n; -\infty \leq a_i \leq b_i \leq \infty, i = 1, \ldots, n\}),$$

where $[a_1, b_1) \times \ldots \times [a_n, b_n) := \emptyset$, if $a_j \geq b_j$ for at least one $j \in \{1, \ldots, n\}$. With

$$\lambda(([a_1, b_1) \times \ldots \times [a_n, b_n)) \cap \mathbb{R}^n) := \begin{cases} \prod_{i=1}^{n} (b_i - a_i) & \text{if } b_i > a_i, i = 1, \ldots, n \\ 0 & \text{otherwise} \end{cases}$$

we obtain a unique measure on $\mathcal{B}^n$. This measure is called *Lebesgue–Borel measure*. The σ-field $\mathcal{B}^n$ is called *Borel* σ-*field on* $\mathbb{R}^n$. Although

$$\mathcal{B}^n \neq \mathcal{P}(\mathbb{R}^n),$$

all important subsets of $\mathbb{R}^n$ (e.g., all open, closed, and compact subsets) are elements of $\mathcal{B}^n$.

Let μ be a measure defined on a σ-field $\mathcal{S}$ on Ω; then each set $A \in \mathcal{S}$ with $\mu(A) = 0$ is called a μ-*null set*. It is obvious that $\mu(B) = 0$ for each subset $B \subseteq A$ of a μ-null set. On the other hand, it is not guaranteed that

$$B \in \mathcal{S} \quad \text{for each} \quad B \subseteq A.$$

A σ-field $\mathcal{S}$ on Ω equipped with a measure $\mu : \mathcal{S} \to \bar{\mathbb{R}}$ is called *complete* if each subset of a μ-null set is an element of $\mathcal{S}$. The σ-field

$$\mathcal{S}_0 := \{A \cup N;\ A \in \mathcal{S},\ N \text{ subset of a } \mu\text{-null set}\}$$

is called *μ-completion of $\mathcal{S}$* with the appropriate measure

$$\mu_0 : \mathcal{S}_0 \to \bar{\mathbb{R}}, \quad (A \cup N) \mapsto \mu(A).$$

Let $\mathcal{S}$ be a σ-field on Ω. The pair $(\Omega, \mathcal{S})$ is called a *measurable space*. With a measure μ on $\mathcal{S}$ the triple $(\Omega, \mathcal{S}, \mu)$ is called a *measure space*.

Let $(\Omega_1, \mathcal{S}_1)$ and $(\Omega_2, \mathcal{S}_2)$ be two measurable spaces. A mapping

$$T : \Omega_1 \to \Omega_2 \quad \text{with} \quad T^{-1}(A') := \{x \in \Omega_1; T(x) \in A'\} \in \mathcal{S}_1 \text{ for all } A' \in \mathcal{S}_2$$

is called *$\mathcal{S}_1$-$\mathcal{S}_2$-measurable*. Consider two measurable spaces $(\Omega_1, \mathcal{S}_1)$ and $(\Omega_2, \mathcal{S}_2)$ with $\mathcal{S}_2 = \sigma(\mathcal{F})$. A mapping $T : \Omega_1 \to \Omega_2$ is $\mathcal{S}_1$-$\mathcal{S}_2$-measurable, iff $T^{-1}(A') \in \mathcal{S}_1$ for all $A' \in \mathcal{F}$.

Let $(\Omega_1, \mathcal{S}_1, \mu_1)$ be a measure space, $(\Omega_2, \mathcal{S}_2)$ be a measurable space, and let $T : \Omega_1 \to \Omega_2$ be $\mathcal{S}_1$-$\mathcal{S}_2$-measurable; then we are able to equip $(\Omega_2, \mathcal{S}_2)$ with a measure μ_2 transformed from $(\Omega_1, \mathcal{S}_1, \mu_1)$ via T by

$$\mu_2 : \mathcal{S}_2 \to \bar{\mathbb{R}}, \quad A' \mapsto \mu_1\left(T^{-1}(A')\right), \quad A' \in \mathcal{S}_2.$$

This measure is called *image measure* of μ_1.

Based on a measurable space $(\Omega, \mathcal{S})$ a $\mathcal{S}$-$\mathcal{B}$-measurable function $e : \Omega \to \mathbb{R}$ is called a *simple function*, if $|e(\Omega)| < \infty$. An interesting class of simple functions is given by indicator functions

$$I_A : \Omega \to \mathbb{R}, \quad \omega \mapsto \begin{cases} 1 & \text{if } \omega \in A \\ 0 & \text{otherwise} \end{cases}, \quad A \in \mathcal{S}.$$

An indicator function I_A indicates whether $\omega \in A$ or not. For each simple function $e : \Omega \to \mathbb{R}$, there exist a natural number n, pairwise disjoint sets $A_1, \ldots, A_n \in \mathcal{S}$, and real numbers $\alpha_1, \ldots, \alpha_n$ with

$$e = \sum_{i=1}^{n} \alpha_i I_{A_i}, \quad \sum_{i=1}^{n} A_i = \Omega.$$

Now, we consider nonnegative simple functions $e : \Omega \to \mathbb{R}_0^+$, $e = \sum_{i=1}^{n} \alpha_i I_{A_i}$, $\alpha_i \geq 0$, $i = 1, \ldots, n$ defined on a measure space $(\Omega, \mathcal{S}, \mu)$, and we define the $(\mu\text{-})$integral for this functions by

$$\int e \, d\mu := \int_{\Omega} e \, d\mu := \sum_{i=1}^{n} \alpha_i \cdot \mu(A_i).$$

It is important to recognize that this integral does not depend on the representation of e by $A_1, \ldots, A_n \in \mathcal{S}$ and $\alpha_1, \ldots, \alpha_n$. Let E be the set of all nonnegative simple functions on $(\Omega, \mathcal{S}, \mu)$; then it is easy to see that

- $\int I_A \, d\mu = \mu(A)$ for all $A \in \mathcal{S}$.
- $\int (\alpha e) d\mu = \alpha \int e \, d\mu$ for all $e \in E$, $\alpha \in \mathbb{R}_0^+$.
- $\int (u + v) d\mu = \int u \, d\mu + \int v \, d\mu$ for all $u, v \in E$.
- If $u(\omega) \leq v(\omega)$ for all $\omega \in \Omega$, then $\int u \, d\mu \leq \int v \, d\mu$, $u, v \in E$.

Consider the σ-field

$$\bar{\mathcal{B}} := \{A \cup B; \ A \in \mathcal{B}, B \subseteq \{-\infty, \infty\}\}$$

on $\bar{\mathbb{R}}$, and let $(\Omega, \mathcal{S})$ be a measurable space and $f : \Omega \to \bar{\mathbb{R}}_0^+$ be a nonnegative $\mathcal{S} - \bar{\mathcal{B}}$-measurable function; then there exists a pointwise monotonic increasing sequence $\{e_n\}_{n \in \mathbb{N}}$ of nonnegative $\mathcal{S}\text{-}\bar{\mathcal{B}}$-measurable simple functions $e_n : \Omega \to \mathbb{R}_0^+$, $n \in \mathbb{N}$, which converges pointwise to f. Therefore, we are able to define the μ-integral for nonnegative $\mathcal{S}\text{-}\bar{\mathcal{B}}$-measurable functions f via

$$\int f \, d\mu := \int_{\Omega} f \, d\mu := \lim_{n \to \infty} \int e_n \, d\mu.$$

Based on a $\mathcal{S}\text{-}\bar{\mathcal{B}}$-measurable function $f : \Omega \to \bar{\mathbb{R}}$, the function

$$f^+ : \Omega \to \bar{\mathbb{R}}_0^+, \quad \omega \mapsto \begin{cases} f(\omega) & \text{if } f(\omega) \geq 0 \\ 0 & \text{otherwise} \end{cases}$$

is called *positive part* of f, and the function

$$f^- : \Omega \to \bar{\mathbb{R}}_0^+, \quad \omega \mapsto \begin{cases} -f(\omega) & \text{if } f(\omega) \leq 0 \\ 0 & \text{otherwise} \end{cases}$$

is called *negative part* of f with the following properties:

- $f^+(\omega) \geq 0$, $f^-(\omega) \geq 0$ for all $\omega \in \Omega$.
- f^+ and f^- are $\mathcal{S}\text{-}\bar{\mathcal{B}}$-measurable.
- $f = f^+ - f^-$.

Using the positive part and the negative part of f, we define

$$\int f \, d\mu := \int_{\Omega} f \, d\mu := \int f^+ \, d\mu - \int f^- \, d\mu,$$

if

$$\int f^+ \, d\mu < \infty \quad \text{or} \quad \int f^- \, d\mu < \infty.$$

Furthermore, we define

$$\int_A f \, d\mu := \int f \cdot I_A \, d\mu.$$

A measure space $(\Omega, \mathcal{S}, \mathbb{P})$ with $\mathbb{P}(\Omega) = 1$ is called *probability space* with probability measure $\mathbb{P}$. For all sets $A \in \mathcal{S}$, the real number $\mathbb{P}(A)$ is called *probability of A*. Let $(\Omega, \mathcal{S}, \mathbb{P})$ be a probability space and let $(\Omega', \mathcal{S}')$ be a measurable space; then a $\mathcal{S}$-$\mathcal{S}'$-measurable function $X : \Omega \to \Omega'$ is called a *random variable*.

If $\Omega' = \mathbb{R}^n$, $n \in \mathbb{N}$, and $\mathcal{S}' = \mathcal{B}^n$, then X is said to be an n-dimensional real random variable. The image measure $\mathbb{P}'$ on $\mathcal{S}'$ is called *distribution of X* and is denoted by $\mathbb{P}_X$. Consider a probability space $(\Omega, \mathcal{S}, \mathbb{P})$ and a random variable $X : \Omega \to \bar{\mathbb{R}}$ with

$$\int X^+ \, d\mathbb{P} < \infty \quad \text{or} \quad \int X^- \, d\mathbb{P} < \infty,$$

then

$$\mathbb{E}(X) := \int X \, d\mathbb{P}$$

is called *expectation of X*. The variance of a random variable $X : \Omega \to \bar{\mathbb{R}}$ with finite expectation $\mathbb{E}(X)$ is given by

$$\mathbb{V}(X) := \int (X - \mathbb{E}(X))^2 \, d\mathbb{P}.$$

Assume that $B \in \mathcal{S}$ and $\mathbb{P}(B) > 0$, then we are able to define another probability measure $\mathbb{P}(\bullet|B)$ (*conditional probability*) on $\mathcal{S}$ by

$$\mathbb{P}(\bullet|B) : \mathcal{S} \to [0, 1], \quad A \mapsto \frac{\mathbb{P}(A \cap B)}{\mathbb{P}(B)}.$$

Consider a measure space $(\Omega, \mathcal{S}, \mu)$ and a $\mathcal{S}$-$\bar{\mathcal{B}}$-measurable function $f : \Omega \to \bar{\mathbb{R}}$ such that

- $f(\omega) \geq 0$ for all $\omega \in \Omega$.
- $\int f \, d\mu = 1$.

Then we obtain a probability measure

$$\mathbb{P} : \mathcal{S} \to [0, 1], \quad A \mapsto \int_A f \, d\mu.$$

The function f is called a *density of $\mathbb{P}$ with respect to μ*.

For the special case $(\mathbb{R}^n, \mathcal{B}^n, \mathbb{P})$, $n \in \mathbb{N}$, the probability measure $\mathbb{P}$ can be expressed by a so-called *distribution function*

$$F : \mathbb{R}^n \to [0, 1], \ (x_1, \ldots, x_n)^\top \mapsto \mathbb{P}\left((-\infty, x_1) \times \ldots \times (-\infty, x_n)\right)$$

The distribution function of an image measure $\mathbb{P}_X$ of an m-dimensional real-valued random variable $X : \mathbb{R}^n \to \mathbb{R}^m$, $m \in \mathbb{N}$, is called *distribution function of X*. The most important class of distribution functions of this type is given by

$$F_{v_{\mathbf{e},\Sigma}} : \mathbb{R}^m \to [0, 1]$$

$$\mathbf{x} \mapsto \int\limits_{(-\infty, x_1)} \cdots \int\limits_{(-\infty, x_m)} \frac{1}{\sqrt{(2\pi)^m \det(\Sigma)}} \cdot \exp\left(-\frac{(\mathbf{x} - \mathbf{e})^\top \Sigma^{-1} (\mathbf{x} - \mathbf{e})}{2}\right) d\mathbf{x}$$

for each $\mathbf{e} \in \mathbb{R}^m$ and each positive definite matrix $\Sigma = (\sigma_{i,j}) \in \mathbb{R}^{m,m}$. A random variable X of this type is called $\mathcal{N}(\mathbf{e}, \Sigma)$ *Gaussian distributed*. Its image measure is given by the Lebesgue density

$$v_{\mathbf{e},\Sigma} : \mathbb{R}^m \to \mathbb{R}, \quad \mathbf{x} \mapsto \frac{1}{\sqrt{(2\pi)^m \det(\Sigma)}} \cdot \exp\left(-\frac{(\mathbf{x} - \mathbf{e})^\top \Sigma^{-1} (\mathbf{x} - \mathbf{e})}{2}\right)$$

with

$$e_i = \mathbb{E}(X_i), \quad i = 1, \ldots, m$$

and

$$\sigma_{i,j} = \mathbb{E}((X_i - e_i)(X_j - e_j)), \quad i, j = 1, \ldots, m.$$

The matrix Σ is called *covariance matrix*.

Based on a probability space $(\Omega, \mathcal{S}, \mathbb{P})$ we have defined a probability measure $\mathbb{P}(\bullet|B)$ (conditional probability) on $\mathcal{S}$ by

$$\mathbb{P}(\bullet|B) : \mathcal{S} \to [0, 1], \quad A \mapsto \frac{\mathbb{P}(A \cap B)}{\mathbb{P}(B)}$$

under the assumption that $\mathbb{P}(B) > 0$. The fact that

$$\mathbb{P}(A|B) = \mathbb{P}(A) \quad \Longleftrightarrow \quad \mathbb{P}(A \cap B) = \mathbb{P}(A) \cdot \mathbb{P}(B)$$

leads to the definition

$$A, B \in \mathcal{S} \quad \text{stochastically independent} \quad \Longleftrightarrow \quad \mathbb{P}(A \cap B) = \mathbb{P}(A) \cdot \mathbb{P}(B).$$

Elements

$$\{A_i \in \mathcal{S}; i \in I\}, I \neq \emptyset$$

of $\mathcal{S}$ are called *stochastically independent*, iff

$$\mathbb{P}\left(\bigcap_{j \in J} A_j\right) = \prod_{j \in J} \mathbb{P}(A_j) \quad \text{for all subsets} \quad J \subseteq I \quad \text{with} \quad 0 < |J| < \infty.$$

A family

$$\{\mathcal{F}_i \subseteq \mathcal{S}; i \in I\}, \quad I \neq \emptyset$$

of subsets of $\mathcal{S}$ is called *stochastically independent*, iff

$$\mathbb{P}\left(\bigcap_{j \in J} A_j\right) = \prod_{j \in J} \mathbb{P}(A_j)$$

for all $A_j \in \mathcal{F}_j$, $j \in J$, and for all $J \subseteq I$ with $0 < |J| < \infty$. Let $(\Omega', \mathcal{S}')$ be a measurable space and let $X : \Omega \to \Omega'$ be a random variable. With $\mathcal{F}$, we denote the set of all σ-fields on Ω such that

$$X \text{ is } \mathcal{C} - \mathcal{S}' - \text{measurable, iff } \mathcal{C} \in \mathcal{F}.$$

The set

$$\sigma(X) := \bigcap_{\mathcal{C} \in \mathcal{F}} \mathcal{C}$$

is called σ-*field generated by* X. It is the smallest σ-field of all σ-fields $\mathcal{A}$ on Ω such that X is $\mathcal{A}$-$\mathcal{S}'$-measurable. The stochastical independence of a family

$$\{X_i : \Omega \to \Omega'; i \in I\}, \quad I \neq \emptyset$$

of random variables is defined by the stochastical independence of

$$\{\sigma(X_i); i \in I\}.$$

A sequence of random variables is a sequence of functions; therefore, one has to introduce different concepts of convergence as done in real calculus also. Let $(\Omega, \mathcal{S}, \mathbb{P})$ be a probability space, let $\{X_i\}_{i \in \mathbb{N}}$ be a sequence of real-valued random variables

$$X_i : \Omega \to \mathbb{R}, \quad i \in \mathbb{N},$$

and let $X : \Omega \to \mathbb{R}$ be a real-valued random variable; then we define for $r \in \mathbb{R}$, $r > 0$:

$$L^r\text{-}\lim_{i \to \infty} X_i = X \quad :\Longleftrightarrow \quad \lim_{i \to \infty} \int |X - X_i|^r \, d\mathbb{P} = 0,$$

where

$$\int |X_i|^r \, d\mathbb{P} < \infty \text{ for all } i \in \mathbb{N} \text{ and } \int |X|^r \, d\mathbb{P} < \infty$$

is assumed.

The sequence $\{X_i\}_{i \in \mathbb{N}}$ converges in $\mathbb{P}$-*measure to* X iff

$$\lim_{i \to \infty} \mathbb{P}\left(\{\omega \in \Omega; \ |X_i(\omega) - X(\omega)| < \epsilon\}\right) = 1$$

for all $\epsilon > 0$, and it converges *almost everywhere to* X iff

$$\mathbb{P}\left(\left\{\omega \in \Omega; \ \lim_{i \to \infty} X_i(\omega) = X(\omega)\right\}\right) = 1.$$

Convergency in distribution is given by

$$\lim_{i \to \infty} \int f \, d\mathbb{P}_{X_i} = \int f \, d\mathbb{P}_X$$

for all infinitely continuously differentiable functions $f : \mathbb{R} \to \mathbb{R}$ with compact support.

A *stochastic process* is a parameterized collection $\{X_t\}_{t \in T}$, $T \neq \emptyset$, of random variables

$$X_t : \Omega \to \mathbb{R}^n$$

based on a probability space $(\Omega, \mathcal{S}, \mathbb{P})$. A function

$$X_\bullet(\omega) : T \to \mathbb{R}^n, \quad t \mapsto X_t(\omega)$$

is called a *path* of $\{X_t\}_{t \in T}$ for each $\omega \in \Omega$. Choose $k \in \mathbb{N}$ and $t_1, \ldots, t_k \in T$; then we find a probability measure on $(\mathbb{R}^k, \mathcal{B}^k)$ via

$$\mathbb{P}_{t_1, \ldots, t_k} : \mathcal{B}^k \to [0, 1],$$

$$(A_1, \ldots, A_k) \mapsto \mathbb{P}(\{\omega \in \Omega; \ X_{t_1}(\omega) \in A_1 \cap \ldots \cap X_{t_k}(\omega) \in A_k\}).$$

Such a probability measure is called a *finite-dimensional distribution of* $\{X_t\}_{t \in T}$. The existence theorem of Kolmogorov (see [Bil86]) describes the existence of a stochastic process using finite-dimensional distributions.

Appendix B
Pseudorandom Numbers

Let $\mathbb{M}$ be a nonempty set with a finite number of elements and let $(\Omega, \mathcal{S}, \mathbb{P})$ be a probability space. Using the measurable space $(\mathbb{M}, \mathcal{P}(\mathbb{M}))$, we consider random variables

$$X_i : \Omega \to \mathbb{M}, \quad i \in \{1, \ldots, |\mathbb{M}|\}$$

with

- $\mathbb{P}_{X_i}(\{x\}) = \frac{1}{|\mathbb{M}|}$ for all $x \in \mathbb{M}$, $i \in \{1, \ldots, |\mathbb{M}|\}$ (the uniform distribution on $\mathbb{M}$).
- The random variables $X_1, \ldots, X_{|\mathbb{M}|}$ are stochastically independent.

Let $\hat{\omega} \in \Omega$ be a result of a random experiment given by $(\Omega, \mathcal{S}, \mathbb{P})$, then we would like to construct a sequence $\{x_n\}_{n \in \mathbb{N}}$ of elements of $\mathbb{M}$ algorithmically with $x_{j \cdot |\mathbb{M}| + i} = X_i(\hat{\omega})$ for all $i = 1, \ldots, |\mathbb{M}|$, $j \in \mathbb{N}_0$; the n-tuple $(x_1, \ldots, x_{|\mathbb{M}|})$ is called a *realization* of $(X_1, \ldots, X_{|\mathbb{M}|})$. For that, we consider a surjective function

$$f : \mathbb{M} \to \mathbb{M}$$

and compute a sequence $\{x_n\}_{n \in \mathbb{N}}$ by

$$x_n = f^{(n-1)}(x_1) := f(f^{(n-2)}(x_1)) \quad \text{with} \quad f^{(0)}(x_1) = x_1,$$

where $x_1 \in \mathbb{M}$ is an arbitrarily chosen starting point, which is called *seed*. Since $|\mathbb{M}| < \infty$, we obtain a periodical sequence $\{x_n\}_{n \in \mathbb{N}}$ with smallest period $s \in \mathbb{N}$. The smallest period of $\{x_n\}_{n \in \mathbb{N}}$ is called *cycle length* of $\{x_n\}_{n \in \mathbb{N}}$. Now, we investigate several choices for $\mathbb{M}$. Consider three integers a, b, and $m > 0$; then we define an equivalence relation $R_m \subseteq \mathbb{Z} \times \mathbb{Z}$ by

$$a \equiv_m b \quad :\Longleftrightarrow \quad (a, b) \in R_m \quad :\Longleftrightarrow$$
$$:\Longleftrightarrow \quad \text{there exists an integer } d \text{ with } a - b = dm.$$

S. Schäffler, *Global Optimization: A Stochastic Approach*, Springer Series in Operations Research and Financial Engineering, DOI 10.1007/978-1-4614-3927-1,
© Springer Science+Business Media New York 2012

If $a \equiv_m b$, then we say *a is congruent to b modulo m*. The integer m is called *modulus* of R_m. For each fixed $m > 0$, each equivalence class (which is called *residue class* also) of R_m has one and only one representative r with

$$0 \le r \le m - 1.$$

The set of residue classes will be denoted by

$$\mathbb{Z}/m\mathbb{Z} := \{[0], \ldots, [m-1]\},$$

where $[i]$ denotes the residue class with $i \in [i]$, $i = 0, \ldots, m-1$. If $a \equiv_m \alpha$ and $b \equiv_m \beta$, then $(a + b) \equiv_m (\alpha + \beta)$ and $ab \equiv_m \alpha\beta$. In other words, we have a well-defined operator

$$\boxplus : \mathbb{Z}/m\mathbb{Z} \times \mathbb{Z}/m\mathbb{Z} \to \mathbb{Z}/m\mathbb{Z},$$

$$([r_1], [r_2]) \mapsto [r_1] \boxplus [r_2] := [r_3] \quad \text{such that} \quad r_1 + r_2 \in [r_3]$$

and a well-defined operator

$$\boxdot : \mathbb{Z}/m\mathbb{Z} \times \mathbb{Z}/m\mathbb{Z} \to \mathbb{Z}/m\mathbb{Z}$$

$$([r_1], [r_2]) \mapsto [r_1] \boxdot [r_2] := [r_4] \quad \text{such that} \quad r_1 \cdot r_2 \in [r_4].$$

The triple $(\mathbb{Z}/m\mathbb{Z}, \boxplus, \boxdot)$ forms a commutative ring.

With $\mathbb{M} = \mathbb{Z}/m\mathbb{Z}$, we choose $[x_1], [a], [b] \in \mathbb{Z}/m\mathbb{Z}$, and we consider the sequence $\{f^{(n-1)}([x_1])\}_{n \in \mathbb{N}}$ given by

$$f : \mathbb{Z}/m\mathbb{Z} \to \mathbb{Z}/m\mathbb{Z}, \quad [x] \mapsto ([a] \boxdot [x]) \boxplus [b].$$

This function is surjective (and thus bijective) if there exists a $[a]^{-1} \in \mathbb{Z}/m\mathbb{Z}$ such that $[a]^{-1} \boxdot [a] = [1]$, which is equivalent to $g.c.d.(a, m) = 1$ (see, e.g., [Kob94]), where the function

$$g.c.d. : \mathbb{Z} \times \mathbb{Z} \to \mathbb{N}$$

computes the **g**reatest **c**ommon **d**ivisor of the arguments. Now, we are looking for stronger conditions on a, b such that $\{f^{(n-1)}([x_1])\}_{n \in \mathbb{N}}$ has the maximum cycle length $s = m$ for all $[x_1] \in \mathbb{Z}/m\mathbb{Z}$. From [Knu97], we know the following result for $m \ge 2$:

Using the function

$$f : \mathbb{Z}/m\mathbb{Z} \to \mathbb{Z}/m\mathbb{Z}, \quad [x] \mapsto ([a] \boxdot [x]) \boxplus [b],$$

the sequence $\{f^{(n-1)}([x_1])\}_{n \in \mathbb{N}}$ has the maximum cycle length $s = m$ for all $[x_1] \in \mathbb{Z}/m\mathbb{Z}$ if the following conditions are fulfilled:

- If any prime number p is a divisor of m, then p is a divisor of $(a - 1)$.

- If 4 is a divisor of m, then 4 is a divisor of $(a-1)$.
- $g.c.d.(b, m) = 1$.

Let $m = 16, a = 9, b = 1$, and $[x_1] = [0]$, for instance; then we obtain the sequence

$$[0], [1], [10], [11], [4], [5], [14], [15], [8], [9], [2], [3], [12], [13], [6], [7], [0], \ldots$$

with maximal cycle length $s = 16$. With the function

$$g : \mathbb{Z}/m\mathbb{Z} \to \mathbb{Z}, \quad [i] \mapsto i,$$

each sequence $\left\{ f^{(n-1)}([x_1]) \right\}_{n \in \mathbb{N}}$ can be transformed to a sequence $\{x_n\}_{n \in \mathbb{N}}$ of real numbers lying in the unit interval by

$$x_n = \frac{g(f^{(n-1)}([x_1]))}{m-1}, \quad n \in \mathbb{N}.$$

The above example leads to

$$0, \frac{1}{15}, \frac{10}{15}, \frac{11}{15}, \frac{4}{15}, \frac{5}{15}, \frac{14}{15}, 1, \frac{8}{15}, \frac{9}{15}, \frac{2}{15}, \frac{3}{15}, \frac{12}{15}, \frac{13}{15}, \frac{6}{15}, \frac{7}{15}, 0, \ldots .$$

Pseudorandom number generators of the form

$$x_n = \frac{g(f^{(n-1)}([x_1]))}{m-1}, \quad n \in \mathbb{N}$$

with

$$f : \mathbb{Z}/m\mathbb{Z} \to \mathbb{Z}/m\mathbb{Z}, \quad [x] \mapsto ([a] \square [x]) \boxplus [b]$$

are called *linear congruential generators*. The quality of these generators depends mainly on the choice of $a, b,$ and m. In [Pre.etal88], the following list of constants is published and organized by the number of necessary bits for the binary representation of $\{g(f^{(n-1)}([x_1]))\}_{n \in \mathbb{N}}$. Each triple (a, b, m) fulfills the above conditions for maximal cycle length, and the corresponding pseudorandom number generator passed several tests of randomness given in [Gen03].

In this book, we need pseudorandom numbers, which approximate realizations of stochastically independent, $\mathcal{N}(0, 1)$ Gaussian distributed random variables. Up to now, we are able to compute pseudorandom numbers, which approximate realizations of stochastically independent, $[0, 1]$-uniformly distributed random variables. Let (u_1, u_2) be a realization of two stochastically independent, $[0, 1]$-uniformly distributed random variables (U_1, U_2) and assume

$$0 < (2u_1 - 1)^2 + (2u_2 - 1)^2 \leq 1,$$

Number of bits	a	b	m
20	106	1,283	6,075
21	211	1,663	7,875
22	421	1,663	7,875
23	430	2,531	11,979
23	936	1,399	6,655
23	1,366	1,283	6,075
24	171	11,213	53,125
24	859	2,531	11,979
24	419	6,173	29,282
24	967	3,041	14,406
25	141	28,411	134,456
25	625	6,571	31,104
25	1,541	2,957	14,000
25	1,741	2,731	12,960
25	1,291	4,621	21,870
25	205	29,573	139,968
26	421	17,117	81,000
26	1,255	6,173	29,282
26	281	28,411	134,456
27	1,093	18,257	86,436
27	421	54,773	259,200
27	1,021	24,631	116,640
27	1,021	25,673	121,500
28	1,277	24,749	117,128
28	741	66,037	312,500
28	2,041	25,673	121,500
29	2,311	25,367	120,050
29	1,807	45,289	214,326
29	1,597	51,749	244,944
29	1,861	49,297	233,280
29	2,661	36,979	175,000
29	4,081	25,673	121,500
29	3,661	30,809	145,800
30	3,877	29,573·	139,968
30	3,613	45,289	214,326
30	1,366	150,889	714,025
31	8,121	28,411	134,456
31	4,561	51,349	243,000
31	7,141	54,773	259,200
32	9,301	49,297	233,280
32	4,096	150,889	714,025
33	2,416	374,441	177,1875
34	17,221	107,839	510,300
34	36,261	66,037	312,500
35	84,589	45,989	217,728

then the tuple (z_1, z_2) given by

$$z_1 := (2u_1 - 1)\sqrt{\frac{-2\ln\left((2u_1 - 1)^2 + (2u_2 - 1)^2\right)}{(2u_1 - 1)^2 + (2u_2 - 1)^2}}$$

$$z_2 := (2u_2 - 1)\sqrt{\frac{-2\ln\left((2u_1 - 1)^2 + (2u_2 - 1)^2\right)}{(2u_1 - 1)^2 + (2u_2 - 1)^2}}$$

is a realization of two stochastically independent, $\mathcal{N}(0, 1)$ Gaussian distributed random variables (Z_1, Z_2) (see [MarBra64]). Consequently, we obtain the following algorithm:

Step 0: (Initialization)
    $i := 1.$
    $j := 1.$
    Choose a, b, m according to the above list such that m is an even number.
    Choose $[x_1] \in \mathbb{Z}/m\mathbb{Z}$.
    goto step 1.

Step 1: (Computation of $[x_2]$)
    compute $[x_2] := ([a] \boxdot [x_1]) \boxplus [b]$.
    goto step 2.

Step 2: ($[0, 1]$-uniformly distributed pseudorandom numbers)
    Compute $u_1 := \frac{g([x_1])}{m-1}$.
    Compute $u_2 := \frac{g([x_2])}{m-1}$.
    goto step 3.

Step 3: $\mathcal{N}(0, 1)$ (Gaussian distributed pseudorandom numbers)
    If $0 < (2u_1 - 1)^2 + (2u_2 - 1)^2 \leq 1$,
        then

$$\text{compute } z_1 := (2u_1 - 1)\sqrt{\frac{-2\ln\left((2u_1 - 1)^2 + (2u_2 - 1)^2\right)}{(2u_1 - 1)^2 + (2u_2 - 1)^2}}.$$

$$\text{compute } z_2 := (2u_2 - 1)\sqrt{\frac{-2\ln\left((2u_1 - 1)^2 + (2u_2 - 1)^2\right)}{(2u_1 - 1)^2 + (2u_2 - 1)^2}}.$$

        $i := i + 2.$
        $j := j + 2.$
        goto step 4.
    else
        $j := j + 2.$
        goto step 4.

Step 4: (Computation of $[x_j], [x_{j+1}]$)

If $j > m$

then

STOP.

else

compute $[x_j] := ([a] \boxdot [x_{j-1}]) \boxplus [b]$.

compute $[x_{j+1}] := ([a] \boxdot [x_j]) \boxplus [b]$.

goto step 5.

Step 5: ($[0, 1]$-uniformly distributed pseudorandom numbers)

Compute $u_j := \frac{g([x_j])}{m-1}$.

Compute $u_{j+1} := \frac{g([x_{j+1}])}{m-1}$.

goto step 6.

Step 6: ($\mathcal{N}(0, 1)$ Gaussian distributed pseudorandom numbers)

If $0 < (2u_j - 1)^2 + (2u_{j+1} - 1)^2 \leq 1$,

then

compute $z_i := (2u_j - 1) \sqrt{\frac{-2\ln\left((2u_j-1)^2+(2u_{j+1}-1)^2\right)}{(2u_j-1)^2+(2u_{j+1}-1)^2}}$.

compute $z_{i+1} := (2u_{j+1} - 1) \sqrt{\frac{-2\ln\left((2u_j-1)^2+(2u_{j+1}-1)^2\right)}{(2u_j-1)^2+(2u_{j+1}-1)^2}}$.

$i := i + 2$.

$j := j + 2$.

goto step 4.

else

$j := j + 2$.

goto step 4.

Appendix C
White Noise Disturbances

In order to analyze optimization problems with white noise disturbances, we have to introduce the concept of generalized functions (see, e.g., [Zem87]). Therefore, we consider the $\mathbb{R}$-vector space K consisting of all infinitely continuously differentiable functions

$$\varphi : \mathbb{R} \to \mathbb{R}$$

with compact support. This vector space is known as the space of all test functions. A sequence $\{\varphi_k\}_{k \in \mathbb{N}}$ of test functions is said to be convergent (with limit $\varphi \in K$) iff

- There exists a compact set $C \subset \mathbb{R}$ such that

$$\varphi_k(x) = 0 \quad \text{for all} \quad k \in \mathbb{N}, \, x \in \mathbb{R} \setminus C,$$

- The sequence $\{\varphi_k\}_{k \in \mathbb{N}}$ and the sequences of the derivatives $\left\{\varphi_k^{(p)}\right\}_{k \in \mathbb{N}}$ of pth order, $p \in \mathbb{N}$, converge uniformly to φ and $\varphi^{(p)}$, respectively.

A generalized function Φ is given by a continuous linear functional

$$\Phi : K \to \mathbb{R}, \quad \varphi \mapsto \Phi(\varphi).$$

The set $\mathcal{D}$ of all generalized functions forms again a $\mathbb{R}$-vector space. Each continuous function $f : \mathbb{R} \to \mathbb{R}$ can be represented by a generalized function

$$\Phi_f : K \to \mathbb{R}, \quad \varphi \mapsto \int_{-\infty}^{\infty} f(t)\varphi(t)dt.$$

The generalized functions

$$\delta_{t_0} : K \to \mathbb{R}, \quad \varphi \mapsto \varphi(t_0), \quad t_0 \in \mathbb{R}$$

S. Schäffler, *Global Optimization: A Stochastic Approach*, Springer Series in Operations Research and Financial Engineering, DOI 10.1007/978-1-4614-3927-1,
© Springer Science+Business Media New York 2012

are called Dirac delta functions. These generalized functions are not representatives of continuous functions because there exists no continuous function $d_{t_0} : \mathbb{R} \to \mathbb{R}$ with

$$\int_{-\infty}^{\infty} d_{t_0}(t)\varphi(t)dt = \varphi(t_0) \quad \text{for all} \quad \varphi \in K.$$

Consider the functions

$$d_{t_0,\sigma^2} : \mathbb{R} \to \mathbb{R}, \quad x \mapsto \frac{1}{\sqrt{2\pi\sigma^2}} \exp\left(-\frac{(x - t_0)^2}{2\sigma^2}\right), \quad \sigma > 0,$$

then we obtain

$$\lim_{\sigma \to 0} \int_{-\infty}^{\infty} d_{t_0,\sigma^2}(t)\varphi(t)dt = \varphi(t_0) \quad \text{for all} \quad \varphi \in K.$$

On the other hand,

$$\lim_{\sigma \to 0} d_{t_0,\sigma^2}(x) = \begin{cases} 0 & \text{if} \quad x \neq t_0 \\ \infty & \text{if} \quad x = t_0 \end{cases}$$

Each generalized function Φ is differentiable, and the derivative of Φ is given by

$$\dot{\Phi} : K \to \mathbb{R}, \quad \varphi \mapsto -\Phi(\varphi'),$$

where φ' denotes the derivative of φ. Let $f : \mathbb{R} \to \mathbb{R}$ be a continuously differentiable function. The fact that

$$\dot{\Phi}_f(\varphi) = -\Phi_f(\varphi') = -\int_{-\infty}^{\infty} f(t)\varphi'(t)dt = \int_{-\infty}^{\infty} f'(t)\varphi(t)dt = \Phi_{f'}(\varphi)$$

makes the definition of $\dot{\Phi}$ reasonable.

Now, we combine the concept of random variables with the concept of generalized functions in the following way (see [Kuo96]):

Let $(\Omega, \mathcal{S}, \mathbb{P})$ be a probability space and let

$$\Phi_\varphi : \Omega \to \mathbb{R}$$

be a real-valued random variable for each $\varphi \in K$. The set

$$\{\Phi_\varphi; \varphi \in K\}$$

of random variables is called a generalized stochastic process, iff the following conditions are fulfilled:

- For all $\alpha, \beta \in \mathbb{R}$ and for all $\varphi, \psi \in K$ holds

$$\Phi_{\alpha\varphi+\beta\psi} = \alpha\Phi_\varphi + \beta\Phi_\psi \quad \mathbb{P} - \text{almost surely.}$$

- Choose $n \in \mathbb{N}$ and consider n sequences of test functions

$$\{\varphi_{1j}\}_{j\in\mathbb{N}}, \ldots, \{\varphi_{nj}\}_{j\in\mathbb{N}}$$

such that

$$\varphi_{ij} \text{ converges to a test function } \varphi_i \text{ with } j \to \infty \text{ for all } i = 1, \ldots, n$$

(convergence in the above scene); then the n-dimensional real-valued random variables $(\Phi_{\varphi_{1j}}, \ldots, \Phi_{\varphi_{nj}})$ converge in distribution to $(\Phi_{\varphi_1}, \ldots, \Phi_{\varphi_n})$.

Let $\{X_t\}_{t\in[0,\infty)}$ be a real-valued stochastic process with continuous paths

$$X_\bullet(\hat{\omega}) : [0, \infty) \to \mathbb{R}, \quad t \mapsto X_t(\hat{\omega}), \quad \hat{\omega} \in \Omega,$$

then this stochastic process can be represented by a generalized stochastic process $\{\Phi_{X,\varphi}; \varphi \in K\}$ with

$$\Phi_{X,\varphi} : \Omega \to \mathbb{R}, \quad \omega \mapsto \int_0^\infty X_t(\omega)\varphi(t)dt.$$

In the following, we consider a one-dimensional Brownian Motion $\{B_t\}_{t\geq 0}$ defined on the probability space $(\Omega, \mathcal{S}, \mathbb{P})$; thus,

1. $\mathbb{P}(\{\omega \in \Omega; B_0(\omega) = 0\}) = 1$.
2. For all $0 \leq t_0 < t_1 < \ldots < t_k, k \in \mathbb{N}$, the random variables

$$B_{t_0}, B_{t_1} - B_{t_0}, \ldots, B_{t_k} - B_{t_{k-1}}$$

are stochastically independent.
3. For every $0 \leq s < t$, the random variable $B_t - B_s$ is $\mathcal{N}(0, (t - s))$ Gaussian distributed.
4. All paths of $\{B_t\}_{t\geq 0}$ are continuous.

The generalized stochastic process representing this Brownian Motion is given by

$$\Phi_{B,\varphi} : \Omega \to \mathbb{R}, \quad \omega \mapsto \int_0^\infty B_t(\omega)\varphi(t)dt.$$

Since the pathwise differentiation $\{\dot{\Phi}_\varphi; \varphi \in K\}$ of a generalized stochastic process $\{\Phi_\varphi; \varphi \in K\}$ yields again a generalized stochastic process, we are able to investigate the generalized stochastic process $\{\dot{\Phi}_{B,\varphi}; \varphi \in K\}$, which has the following properties:

- $\mathbb{E}(\dot{\Phi}_{B,\varphi}) = 0$ for all $\varphi \in K$.
- $\mathrm{Cov}(\dot{\Phi}_{B,\varphi}, \dot{\Phi}_{B,\psi}) = \int\limits_{-\infty}^{\infty} \varphi(t)\psi(t)dt$.
- For arbitrarily chosen linearly independent functions $\varphi_1, \ldots, \varphi_n \in K, n \in \mathbb{N}$, the random variable $(\dot{\Phi}_{B,\varphi_1}, \ldots, \dot{\Phi}_{B,\varphi_n})$ is Gaussian distributed.

If one tries to describe the covariance function $\mathrm{Cov}(\dot{\Phi}_{B,\varphi}, \dot{\Phi}_{B,\psi})$ as a classical function in two variables t and s, then this function has to behave like

$$\lim_{\sigma \to 0} d_{s,\sigma^2}(t) \left(= \begin{cases} 0 & \text{if} \quad s \neq t \\ \infty & \text{if} \quad s = t \end{cases} \right).$$

For this reason and because

$$\mathbb{E}(\dot{\Phi}_{B,\varphi}) = 0 \quad \text{for all} \quad \varphi \in K,$$

the generalized stochastic process $\{\dot{\Phi}_{B,\varphi}; \varphi \in K\}$ serves as a typical random noise process in engineering and is called a white noise process. Although this process is not a representative of a classical stochastic process, it is often described by

$$\nu_t : \Omega \to \mathbb{R}, \quad t \in [0, \infty)$$

in the literature. Using an n-dimensional Brownian Motion

$$\{\mathbf{B}_t\}_{t \in [0,\infty)} = \begin{pmatrix} \{B_{1,t}\}_{t \in [0,\infty)} \\ \vdots \\ \{B_{n,t}\}_{t \in [0,\infty)} \end{pmatrix},$$

then $\{B_{1,t}\}_{t \in [0,\infty)}, \ldots, \{B_{n,t}\}_{t \in [0,\infty)}$ are stochastically independent one-dimensional Brownian Motions, and we obtain an n-dimensional white noise process

$$\left\{ \begin{pmatrix} \dot{\Phi}_{B_1,\varphi} \\ \vdots \\ \dot{\Phi}_{B_n,\varphi} \end{pmatrix} ; \varphi \in K \right\}$$

incorrectly denoted by $\{\mathbf{N}_t\}_{t \in [0,\infty)}$ in the literature.

Let

$$\mathbf{Y}_t : \Omega \to \mathbb{R}^{n,n}, \quad t \in [0, \infty),$$

be a matrix-valued stochastic process defined on $(\Omega, \mathcal{S}, \mathbb{P})$ with continuous paths; then the Fisk–Stratonovich integral of $\{\mathbf{Y}_t\}_{t \in [0,\infty)}$ with respect to the Brownian Motion $\{\mathbf{B}_t\}_{t \in [0,\infty)}$ was defined by

$$\int_0^T \mathbf{Y}_t \circ d\mathbf{B}_t := L^2\text{-}\lim_{q \to \infty} \sum_{i=1}^{p_q} \mathbf{Y}_{\frac{t_i + t_{i-1}}{2}} (\mathbf{B}_{t_i} - \mathbf{B}_{t_{i-1}}).$$

Substituting

$$\sum_{i=1}^{p_q} \mathbf{Y}_{\frac{t_i + t_{i-1}}{2}} (\mathbf{B}_{t_i} - \mathbf{B}_{t_{i-1}}) \text{ by } \sum_{i=1}^{p_q} \mathbf{Y}_{\frac{t_i + t_{i-1}}{2}} \frac{\mathbf{B}_{t_i} - \mathbf{B}_{t_{i-1}}}{t_i - t_{i-1}} (t_i - t_{i-1}),$$

we define

$$\int_0^T \mathbf{Y}_t \mathbf{N}_t \, dt := \int_0^T \mathbf{Y}_t \circ d\mathbf{B}_t.$$

In several applications like image processing and pattern recognition, one has to minimize a twice continuously differentiable objective function $f : \mathbb{R}^n \to \mathbb{R}$, which is not known explicitly, but informations about the gradient of f are given in the following way:

For each continuous curve

$$k : [0, \infty) \to \mathbb{R}^n,$$

the evaluation of the gradients of f along these curves is connected with additive white noise disturbances:

$$\nabla f(k(t)) + \mathbf{N}_t, \quad \text{instead of} \quad \nabla f(k(t)), \quad t \geq 0,$$

or, more correctly in terms of generalized functions:

$$\Phi_{\nabla f \circ k}(\varphi) + \begin{pmatrix} \dot{\Phi}_{B_1, \varphi} \\ \vdots \\ \dot{\Phi}_{B_n, \varphi} \end{pmatrix} \quad \text{instead of} \quad \Phi_{\nabla f \circ k}(\varphi), \quad \varphi \in K,$$

which means:

$$\begin{pmatrix} \int_0^\infty (\nabla f(k(t))_1 \varphi(t) - B_{1,t} \varphi'(t)) dt \\ \vdots \\ \int_0^\infty (\nabla f(k(t))_n \varphi(t) - B_{n,t} \varphi'(t)) dt \end{pmatrix} \quad \text{instead of} \quad \begin{pmatrix} \int_0^\infty \nabla f(k(t))_1 \varphi(t) dt \\ \vdots \\ \int_0^\infty \nabla f(k(t))_n \varphi(t) dt \end{pmatrix},$$

$\varphi \in K$.

Now, we have to deal with a comparison of function values of f. Assume that we have computed two points $\mathbf{x}_1, \mathbf{x}_2 \in \mathbb{R}^n$. With the line

$$\gamma : [0, 1] \to \mathbb{R}^n, \quad t \mapsto \mathbf{x}_1 + t(\mathbf{x}_2 - \mathbf{x}_1),$$

we obtain:

$$f(\mathbf{x}_2) - f(\mathbf{x}_1) = \int_0^1 \dot{f}(\gamma(t))dt = \int_0^1 \nabla f(\gamma(t))^\top (\mathbf{x}_2 - \mathbf{x}_1)dt$$

$$= \left(\int_0^1 \nabla f(\gamma(t))dt \right)^\top (\mathbf{x}_2 - \mathbf{x}_1).$$

Since we can evaluate only realizations of

$$\nabla f(\gamma(t)) + \mathbf{N}_t \quad \text{instead of} \quad \nabla f(\gamma(t)),$$

we are only able to compute numerically a realization of the random variable

$$\Delta f : \Omega \to \mathbb{R}, \quad \omega \mapsto \int_0^1 (\nabla f(\gamma(t)) + \mathbf{N}_t(\omega))^\top (\mathbf{x}_2 - \mathbf{x}_1)dt$$

$$= \left(\int_0^1 (\nabla f(\gamma(t)) + \mathbf{N}_t(\omega))dt \right)^\top (\mathbf{x}_2 - \mathbf{x}_1)$$

$$= \left(\int_0^1 \nabla f(\gamma(t))dt \right)^\top (\mathbf{x}_2 - \mathbf{x}_1) + \left(\int_0^1 \mathbf{N}_t(\omega)dt \right)^\top (\mathbf{x}_2 - \mathbf{x}_1)$$

$$= f(\mathbf{x}_2) - f(\mathbf{x}_1) + (\mathbf{B}_1(\omega) - \mathbf{B}_0(\omega))^\top (\mathbf{x}_2 - \mathbf{x}_1).$$

Therefore, Δf is a $\mathcal{N}\left(f(\mathbf{x}_2) - f(\mathbf{x}_1), \|\mathbf{x}_2 - \mathbf{x}_1\|_2^2 \right)$ Gaussian distributed random variable. If the computed realization of Δf is greater than zero, then we conclude $f(\mathbf{x}_2) > f(\mathbf{x}_1)$ and vice versa. The error probability is given by

$$\int_{-\infty}^{0} \frac{1}{\sqrt{2\pi \|\mathbf{x}_2 - \mathbf{x}_1\|_2^2}} \exp\left(-\frac{(x - |f(\mathbf{x}_2) - f(\mathbf{x}_1)|)^2}{2\|\mathbf{x}_2 - \mathbf{x}_1\|_2^2}\right) dx$$

$$= \Phi\left(-\frac{|f(\mathbf{x}_2) - f(\mathbf{x}_1)|}{\|\mathbf{x}_2 - \mathbf{x}_1\|_2}\right),$$

where Φ denotes the standard normal cumulative distribution function. In [Mör93], numerical examples of this approach are summarized with up to $n = 5$ variables.

It is important to distinguish between the Brownian Motion $\{\mathbf{B}_t\}_{t \geq 0}$ implicitly given by the white noise process and a Brownian Motion $\{\bar{\mathbf{B}}_t\}_{t \geq 0}$ used for randomization of the curve of steepest descent. Both processes can be assumed to be stochastically independent. Since

$$\mathbf{X}(t, \omega) = \mathbf{x}_0 - \int_0^t (\nabla f(\mathbf{X}(\tau, \omega)) + \mathbf{N}_\tau(\omega)) d\tau + \epsilon \left(\bar{\mathbf{B}}_t(\omega) - \bar{\mathbf{B}}_0(\omega)\right)$$

$$= \mathbf{x}_0 - \int_0^t (\nabla f(\mathbf{X}(\tau, \omega)) + \mathbf{N}_\tau(\omega)) d\tau + \epsilon \int_0^t \mathbf{I}_n \circ d\bar{\mathbf{B}}_\tau(\omega)$$

$$= \mathbf{x}_0 - \int_0^t \nabla f(\mathbf{X}(\tau, \omega)) d\tau - \int_0^t \mathbf{N}_\tau(\omega) d\tau + \epsilon \int_0^t \bar{\mathbf{N}}_\tau(\omega) d\tau$$

$$= \mathbf{x}_0 - \int_0^t \nabla f(\mathbf{X}(\tau, \omega)) d\tau + \int_0^t \underbrace{(\epsilon \bar{\mathbf{N}}_\tau(\omega) - \mathbf{N}_\tau(\omega))}_{\sqrt{\epsilon^2 + 1} \cdot \dot{\bar{\mathbf{B}}}_\tau(\omega)} d\tau$$

$$= \mathbf{x}_0 - \int_0^t \nabla f(\mathbf{X}(\tau, \omega)) d\tau + \sqrt{\epsilon^2 + 1} \left(\tilde{\mathbf{B}}_t(\omega) - \tilde{\mathbf{B}}_0(\omega)\right),$$

the theory of section three is applicable with $\sqrt{\epsilon^2 + 1}$ instead of ϵ.

References

[Al-Pe.etal85] Allufi-Pentini, F., Parisi, V., Zirilli, F.: Global optimization and stochastic differential equations. JOTA **47**, 1–16 (1985)

[Bar97] Barnerssoi, L.: Eine stochastische Methode zur globalen Minimierung nichtkonvexer Zielfunktionen unter Verwendung spezieller Gradientenschätzungen. Shaker, Aachen (1997)

[BesRit88] Best, M.J., Ritter, K.: A quadratic programming algorithm. ZOR **32**, 271–297 (1988)

[Bil86] Billingsley, P.: Probability and Measure. Wiley, New York (1986)

[Bul.etal03] Bulger, D., Baritompa, W.P., Wood, G.R.: Implementing pure adaptive search with Grover's quantum algorithm. JOTA **116**, 517–529 (2003)

[Chi.etal87] Chiang, T., Hwang, C., Sheu, S.: Diffusions for global optimization in $\mathbb{R}^n$. SIAM J. Contr. Optim. **25**, 737–753 (1987)

[Conn.etal00] Conn, A.R., Gould, N.I.M., Toint, P.L.: Trust-Region Methods. MPS-SIAM series on optimization, Philadelphia (2000)

[Cottle.etal92] Cottle, R.W., Pang, J.S., Stone, R.E.: The Linear Complementarity Problem. Academic, San Diego (1992)

[Crank84] Crank, J.: Free and Moving Boundary Problems. Clarendon Press, Oxford (1984)

[Elw82] Elworthy, K.D.: Stochastic Differential Equations on Manifolds. Cambridge University Press, Cambridge (1982)

[Flo00] Floudas, C.A.: Deterministic Global Optimization. Kluwer, Dordrecht (2000)

[Flo.etal99] Floudas, C.A., Pardalos, P.M., Adjiman, C.S., Esposito, W.R., Gümüs, Z.H., Harding, S.T., Klepeis, J.L., Meyer, C.A., Schweiger, C.A.: Handbook of Test Problems in Local and Global Optimization. Kluwer, Dordrecht (1999)

[Fried06] Friedman, A.: Stochastic Differential Equations and Applications. Dover, New York (2006)

[GemHwa86] Geman, S., Hwang, C.: Diffusions for global optimization. SIAM J. Contr. Optim. **24**, 1031–1043 (1986)

[Gen03] Gentle, J.E.: Random Number Generation and Monte Carlo Methods. Springer, Berlin (2003)

[Hajek88] Hajek, B.: Cooling schedules for optimal annealing. Math. Oper. Res. **13**, 311–329 (1988)

[Has.etal05] Hashimoto, K., Kobayashi, K., Nakao, M: Numerical verification methods for solutions of the free boundary problem. Numer. Funct. Anal. Optim. **26**, 523–542 (2005)

[Has80] Hasminskij, R.Z.: Stochastic Stability of Differential Equations. SIJTHOFF & NOORDHOFF, Amsterdam (1980)

S. Schäffler, *Global Optimization: A Stochastic Approach*, Springer Series in Operations Research and Financial Engineering, DOI 10.1007/978-1-4614-3927-1,
© Springer Science+Business Media New York 2012

[HenTót10]   Hendrix, E.M.T., Tóth, B.: Introduction to Nonlinear and Global Optimization. Springer, Berlin (2010)

[Holl75]   Holland, J.H.: Adaption in Natural and Artificial Systems. University of Michigan Press, MI (1975)

[HorTui96]   Horst, R., Tui, H.: Global Optimization: Deterministic Approaches. Springer, Berlin (1996)

[Jahn10]   Jahn, J.: Vector Optimization. Springer, Berlin (2010)

[KarShr08]   Karatzas, I., Shreve, S.E.: Brownian Motion and Stochastic Calculus. Springer, Berlin (2008)

[Kar63]   Karnopp, D.C.: Random search techniques for optimization problems. Automata 1, 111–121 (1963)

[KenEber95]   Kennedy, J., Eberhart, R.C.: Particle swarm optimization. In: Proceedings of IEEE International Conference on Neural Networks, pp. 1942–1948, Piscataway, NJ (1995)

[Knu97]   Knuth, D.E.: The Art of Computer Programming, vol. 2: Seminumerical Algorithms. Addison-Wesley, MA (1997)

[Kob94]   Koblitz, N.: A Course in Number Theory and Cyrptography. Springer, Berlin (1994)

[Kuo96]   Kuo, H-H.: White Noise Distribution Theory. CRC Press, Boca Raton (1996)

[LaaAarts87]   van Laarhoven, P.J.M., Aarts, E.H.L.: Simulated Annealing: Theory and Applications. D. Reidel Publishing Co., Dordrecht (1987)

[vanLint98]   van Lint, J.H.: Introduction to Coding Theory. Springer, Berlin (1998)

[Lev44]   Levenberg, K.: A method for the solution of certain nonlinear problems in least squares. Q. Appl. Math. 2, 164–168 (1944)

[Lue84]   Luenberger, D.G.: Linear and Nonlinear Programming. Addison-Wesley, MA (1984)

[MarBra64]   Marsaglia, G., Bray, T.A.: A convenient method for generating normal variables. SIAM Rev. 6, 260–264 (1964)

[McShane73]   McShane, E.J.: The Lagrange multiplier rule. Am. Math. Mon. 8, 922–925 (1973)

[Met.etal53]   Metropolis, G., Rosenbluth, A., Rosenbluth, M., Teller, A., Teller, E.: Equation for state calculations by fast computing machines. J. Chem. Phys. 21, 1087–1092 (1953)

[Mitra.etal86] Mitra, D., Romeo, F., Sangiovanni-Vincentelli, A.: Convergence and finite-time behavior of simulated annealing. SIAM J. Contr. Optim. 18, 747–771 (1986)

[Mör93]   Mörtlbauer, W.: Globale stochastische Optimierung unter Verwendung stochastischer Integralgleichungen. Thesis, TU München, Munich (1993)

[Owen68]   Owen, G.: Game Theory. W.B. Saunders Company, London (1968)

[Pat.etal89]   Patel, N.R., Smith, R.L., Zabinsky, Z.B.: Pure adaptive search in Monte Carlo optimization. Math. Program. 43, 317–328 (1989)

[Pin70]   Pincus, M.: A Monte Carlo method for the approximate solution of certain types of constrained optimization problems. Oper. Res. 18, 1225–1228 (1970)

[Pre.etal88]   Press, W.H., Flannery, B.P., Teukolsky, S.A., Vetterling, W.T.: Numerical recipes in C: The Art of Scientific Computing. Cambridge University Press, Cambridge (1988)

[Proa95]   Proakis, J.G.: Digital Communications. McGraw-Hill, New York (1995)

[Pro95]   Protter, P.: Stochastic Integration and Differential Equations. A New Approach. Springer, Berlin (1995)

[RenRog04]   Renardy, M., Rogers, R.C.: An Introduction to Partial Differential Equations. Springer, Berlin (2004)

[RitSch94]   Ritter, K., Schäffler, S.: A stochastic method for constrained global optimization. SIAM J. Optim. 4, 894–904 (1994)

[Sal.etal02]   Salamon, P., Sibani, P., Frost, R.: Facts, Conjectures, and Improvements for Simulated Annealing. SIAM, Philadelphia (2002)

[Schäfer08]	Schäfer, U.: Das lineare Komplementaritätsproblem. Springer, Berlin (2008)
[Schä93]	Schäffler, S.: Global Optimization Using Stochastic Integral Equations. Habilitation Thesis, TU München (1993)
[Schä95]	Schäffler, S.: Global Optimization Using Stochastic Integration. Roderer, Regensburg (1995)
[Schä97]	Schäffler, S.: Decodierung binärer linearer Blockcodes durch globale Optimierung. Roderer, Regensburg (1997)
[SchäWar90]	Schäffler, S., Warsitz, H.: A trajectory-following method for unconstrained optimization. JOTA **67**, 133–140 (1990)
[Schä.etal02]	Schäffler, S., Schultz, R., Weinzierl, K.: Stochastic method for the solution of unconstrained vector optimization problems. JOTA **114**(1), 209–222 (2002)
[Schn11]	Schneider, E.: Ein SQP-Verfahren zur globalen Optimierung. Shaker, Aachen (2011)
[Stö00]	Stöhr, A.: A Constrained Global Optimization Method Using Stochastic Differential Equations on Manifolds. Roderer, Regensburg (2000)
[StrSer00]	Strongin, R., Sergeyev, Y.: Global Optimization with Non-Convex Constraints. Kluwer, Dordrecht (2000)
[Stu03]	Sturm, T.F.: Stochastische Analysen und Algorithmen zur Soft-Decodierung binärer linearer Blockcodes. Thesis, Universität der Bundeswehr München (2003)
[Tho78]	Thorpe, J.A.: Elementary Topics in Differential Geometry. Springer, Berlin (1978)
[WoodZab02]	Wood, G.R., Zabinsky, Z.B.: Stochastic adaptive search. In: Pardalos, P., Romeijn, E. (eds.) Handbook of Global Optimization, vol. 2, pp. 231–249. Kluwer, Dordrecht (2002)
[Zab03]	Zabinsky, Z.B.: Stochastic Adaptive Search for Global Optimization. Kluwer, Dordrecht (2003)
[Zem87]	Zemanian, A.H.: Distribution Theory and Transform Analysis: An Introduction to Generalized Functions, with Applications. Dover, New York (1987)
[Zhi91]	Zhigljavsky, A.: Theory of Global Random Search. Kluwer, Dordrecht (1991)
[ZhiŽil08]	Zhigljavsky, A., Žilinskas, A.: Stochastic Global Optimization. Springer, Berlin (2008)

Index

S. Schäffler, *Global Optimization: A Stochastic Approach*, Springer Series in Operations   145
Research and Financial Engineering, DOI 10.1007/978-1-4614-3927-1,
© Springer Science+Business Media New York 2012